WHAT YOU NEED TO KNOW TO LIVE WITH CHEMICALS

by

Ralph I. Freudenthal, Ph.D.

and

Susan Loy Freudenthal

Hill and Garnett Publishing, Inc.
Greens Farms, Connecticut

Published by Hill and Garnett Publishing, Inc., P.O. Box 180
Greens Farms, CT 06436.

Cover design by Jenniffer Lee

Figures 2.1 and 7.2 by Holly Nichols

Library of Congress Catalog Card Number 89-85145

ISBN 0-9623813-0-6

Printed in the United States of America

Preface

As a working scientist, I am often asked the following questions by non-scientist friends:

"Isn't a *natural* food always safe?"

"Isn't a substance safe to use if the government allows it to be sold?"

"What does *risk* really mean when we are talking about cancer-causing chemicals? How is risk determined? By whom?"

"Aren't man-made chemicals always harmful to us?"

"I can't do anything important to stop pollution, can I?"

These questions are straight-forward and seem simple enough. Actually, the answers are sometimes rather complicated, and unfortunately, in addition, those asking the questions often do not have a sufficient background knowledge of science to enable them to relate my answers to the overall picture without considerable discussion. Many of these friends are quite well educated, but in our age of specialization, it is not always easy for a specialist in one field of endeavor to acquire a good working knowledge of another field which may not be closely related. Consequently, it is often difficult for "experts" in different fields to converse meaningfully about one another's work.

In my early career as a scientist, I was employed by non-profit research institutes. I joined the ranks of those searching for "the cure for cancer," and I knew that as a cancer research scientist, I was participating in a noble profession. After several years, my interests expanded from cancer research to the broader study of the causation of toxicity. Later, I joined a well-known chemical company to establish a state-of-the-art toxicology testing laboratory, with money not much of an obstacle. This opportunity represented the "dream come true" for this research scientist. It provided an opportunity to design and operate a first-rate laboratory devoted to testing chemicals to determine their safety for both man and his environment.

I set about my task with great excitement, but I soon discovered that many people regard research, when associated with the chemical industry, as some sort of "dirty work," even though all health researchers use essentially the same technology and procedures toward the same ultimate goal of protecting human and environmental health. I was perplexed by the sometimes hostile reactions of the public toward chemical industry scientists, and I was convinced that it was mostly due to a misperception on their part concerning our motives. I am aware that in the past the chemical industry has been guilty of wrongdoing, but with recent legislation (beginning about 50 years ago), the industry is now compelled to develop products which promote human and environmental health. However, the assurance of human and environmental well-being involves the cooperative efforts of the entire world community—all industries and businesses, governments, and individuals.

This book is written for the non-scientist who wants to acquire sufficient knowledge of biology and chemistry to assess for himself (1) the importance of the much-publicized environmental and human health issues which challenge today's world, (2) the causes and relationships of the issues, and (3) the validity of various proposed solutions.

Our general lack of understanding of the basic issues affecting our health and our environment is partly the result of our reliance on politicians and special interest groups to make important decisions for us. We often don't question their decisions or take a truly personal interest in exploring the possible consequences. Although we devote significant time and effort to making an informed choice of automobile, VCR, washing machine, or vacation, most of us don't expend nearly enough care when it comes to making choices of many other products which affect our lives.

It is the belief of my co-author and I that only an informed citizenry can effect the needed changes in our society's system of chemical use. It is at the level of the individual that product demand is created and it is also at the individual level that a product's worst consequences are experienced, usually as adverse health effects. As a population, we are generally concerned about the health and safety of ourselves and our environment. As individuals, it is our responsibility to gather all the needed facts and then begin working together for a safer, healthier life for all.

We must begin by seeking to understand both the "benefits" and the "risks" involved in the use of a given chemical substance. When evaluating risks, we should consider the effects of the manufacturing process as well as the ultimate disposal procedure. Perhaps the risks to the environment caused by a product's manufacture and disposal outweigh the benefits of its use. If so, a suitable alternative can often be found. For example, does the convenience provided by plastic food containers justify the long-term pollution created by both their manufacture and disposal? Can we not get along with using paper containers instead? To answer, we must evaluate many factors relating to the use of each.

Over the past several years, I have made non-technical presentations to service organizations, such as the Rotary Club, the Chamber of Commerce, and the League of Women Voters, in which I discuss the impact of chemicals on our lives and on the environment. Topics include: chemicals in our foods, hazards associated with pesticide use, risks associated with "natural" foods, hazardous waste, the contamination of our environment, and the concept of "relative risks." At the conclusion of these discussions, the same question always emerges–"Is there a book available, written in simple language, which objectively describes these important issues?" There is none, and this book responds to that need.

Ralph I. Freudenthal, Ph.D.
April 20, 1989

Introduction

This book is intended to provide a brief overview of some of the many important environmental and human health issues that challenge today's world and its inhabitants. The book seeks to provide a sense of how each issue relates to the others as well as to the whole.

We have attempted to present as simply as possible the basic facts necessary for a thoughtful consideration of the issues. A Glossary of some perhaps unfamiliar technical terms is provided.

Some conclusions and proposed solutions are ours but are shared by many of our peers. It is hoped that the reader will go on to a more thorough study of these topics that we can discuss only briefly in this volume.

Ralph I. Freudenthal and Susan L. Freudenthal

Table of Contents

Confusion About Chemicals

A swift, negative reaction is experienced by many people upon hearing the word *chemical*. Chemicals are usually considered to be dangerous materials which are produced by, used by, and disposed of by the chemical industry. Chemicals are blamed for causing cancer, birth defects, environmental pollution, and most of the other ills facing the world today. In short, chemicals are held responsible for the general deterioration of our earth and all its inhabitants.

Media headlines and exposés often exclaim the dangers related to the use of pesticides, drugs, household cleaners, and other chemicals in daily use. We are told that our homes may be unsafe because of formaldehyde foam insulation and radon gas, that many foods are unsafe due to the presence of cancer-causing pesticides, and that some drinking water is unsafe due to the presence of harmful amounts of chemicals and pesticides. With such health issues being brought to our attention nearly every day, it becomes obvious that there is much we must understand if we are to be able to decide for ourselves if we really are at risk.

We sometimes feel confused by seemingly endless reports of impending disaster. We are often concerned that we are not being given sufficient information to allow us to make our own judgments. We fear that we and our environment may not be receiving the protection from harmful materials expected under the laws of our government. We feel frustration with a news media which grabs our attention with allegations which often are later found to be untrue, unsupported by the facts. We are unsure what is meant by "risk" and also how it is determined and by whom.

Somewhere along the way we have gotten the idea that we can live in a risk-free world. This has never been so and never will be. We accept a great many risks in our daily living, such as traveling in automobiles and airplanes, sailing and skiing, etc. We ignore the apparently acceptable risks in such activities while focusing on our desire to achieve totally risk-free products, particularly those we call "chemical products."

Natural Chemicals and Man-Made Chemicals

We and the whole world of matter surrounding us are made up entirely of chemicals. There exist more than 6,000,000 different chemicals which have been registered and written about in recent scientific literature. Some chemicals are man-made, but the majority are of natural origin.

We have the mistaken idea that all natural chemicals are safe and all man-made chemicals are harmful. This is not so. A great many man-made products are entirely safe, and a great many natural products are quite deadly. Most of the man-made chemicals to which we are exposed are safe for their intended uses. Many natural chemicals are highly toxic, causing cancer, birth defects, and other adverse health effects. There are serious problems associated with some so-called "health foods."

Some activists suggest that man-made chemicals are the cause of all human and environmental problems and therefore those chemicals should be banned. If we suppose that these activists are correct, then we must consider whether we *could* ban all man-made chemicals, and further, whether we *would*, or *should*. Man-made (synthetic) chemicals include: most medicines and other pharmaceuticals; synthetic fibers and plastics used in clothing, in homes, and in automobiles; synthetic rubber for tires; household cleaners such as detergents, stain removers, disinfectants, bathroom cleaners, and mildewicides; shampoos, soaps, and other cosmetics; color dyes used in clothing, upholstery, and paints; plastic products such as food containers, garbage bags, computer components, automobile parts, toys, and telephones; fiberglass; insecticides and other pesticides; and dry cleaning solvents.

Imagine trying to maintain our standard of living without such chemicals. There would be no automobiles, no airplanes, no computers, no formica counter tops, only natural fiber clothing and carpets colored with berry juices and bark extracts, no plastics, very few pharmaceuticals at considerably higher prices, and a severe shortage of fresh seasonal farm produce with any such foods demanding exceptionally high prices. Obviously, there is much we probably wouldn't want to give up, but we do want some means of assurance that the chemical products reaching the marketplace are safe for us and our environment.

Reason and Caution Are Needed

There is a critical need to direct scientific ingenuity and advancements in technology towards methods and products that are more environmentally acceptable. Industry, national and state governments, international organizations, and special interest groups can work together to assure that new products introduced into our environment have a neutral or positive environmental impact, have global value, and provide a benefit to mankind. New technologies must be carefully directed along lines which will not create additional unreasonable risks or further damage our ecosystems.

All things are toxic in large enough amounts. Analytical testing techniques have become so sensitive that it is possible to find contaminants in almost any material. While it is appropriate to determine the risk involved in the use of a given substance, we must then weigh the risk against the benefit before we decide whether or not to use the substance. Our efforts should be directed toward an overall balance, with equity for all, worldwide.

In principle, we commit to a pollution-free environment; however, practically, we soon discover that such an ideal is unattainable. We demand that materials which do not biodegrade in the ground not be buried, and yet we are continually adding to the stockpile of nonbiodegradable wastes needing burial sites. For convenience, we use plastic trash bags to contain our garbage, leaves, and other discards. These plastic garbage bags will remain in the earth, unchanged, for thousands of years, as will the plastic milk and margarine and cola containers, plastic food wrap, garment covers, cups, and spoons.

Compromises are necessary, and these compromises are often based on subjective judgments. Most of us are willing to practice a responsible program of wise use balanced with reasonable conservation in order to safeguard ourselves and our environment. This would require strong personal principles and some practical constraints on the use of everyday conveniences, but these compromises could make a significant difference.

Modern technology has increased our lifespan and provided many valuable and useful tools for convenient living. We need not succumb to the extremists who would have us dump it all and go back to what they envision as "risk-free natural living." After we take an objective look, we will surely conclude that natural living is definitely not risk-free.

If we reject the simplistic notion that society can eliminate all man-made chemicals, then we must educate ourselves to the realization that small amounts of most chemicals are essentially harmless, and we must endeavor to protect humans and the environment from excessive exposure to all hazardous materials, man-made or natural.

Living *is* risking; survival depends upon successful adaptation to life's ever-changing conditions. In order to make informed and equitable decisions on issues affecting our health and environment, we must first acquire a basic understanding of the facts on which these decisions must be based.

Various laws have been passed in an effort to safeguard both man and the environment. Beginning in 1906 with the Pure Food and Drug Act, our government has passed numerous laws in an effort to control the release of toxic chemicals into our environment and to protect the consumer from potentially hazardous substances. This broad coverage includes regulation of: exposure to toxic chemicals in the workplace; the manufacture and use of pesticides; pre-manufacture testing of new chemicals; the transportation, storage, and disposal of hazardous wastes; and the cleanup of hazardous waste sites. Certainly, these comprehensive laws should be able to provide us with sufficient protection from harmful chemicals. However, while numerous effective laws exist, their enforcement is inconsistent and results in overregulation of some materials and underregulation of others.

Certain types of chemicals – pharmaceuticals, pesticides, and food additives – have received a great deal of adverse publicity in recent years. These chemicals have undergone careful scrutiny using very sensitive testing procedures. However, the same intense level of thorough investigation has not been applied to other chemical products in daily use, such as cosmetics, dyes and colorants, and "natural" health products.

The level of testing for many health and beauty products is often quite superficial, and these products encounter less regulation by the government. Shouldn't a consumer be entitled by law to the same level of protection for all the products he uses? What is the value in overprotecting us from some while hardly protecting us at all from others?

We, the general public, can no longer leave it to the experts and lobbyists to decide for us what is good for us. We must educate ourselves and learn to "speak their language" in order to participate in the solution to these vital issues which challenge our future. In order to safeguard our environment, we need to think in broad terms as to how our individual and collective efforts can impact global issues.

The late René Dubos, an internationally recognized expert on environmental issues and the author of *Only One Earth*, challenged us to "Think globally, and act locally." In his many books, he points out the need to have a balance between industrial development and the conservation of natural resources. Unfortunately, many industries and governments see the world as a treasure to be harvested bare and sold for greatest gain. They view the world as an inexhaustible source of commodities, raw materials, and exploitable land. We—the employees of these industries, the constituents of these governments—have both the responsibility and the need to preserve our natural environment.

The preservation of our environment is our gift *and* our obligation to future generations. However, we must go forward from our situation in the present. We can not go back to some perfect condition which never really existed except in the fantasies of extremists.

Chemicals, Chemicals, Chemicals

We are currently faced with serious health and environmental problems resulting from the use of chemicals, both natural and man-made. Chemicals are cited for depleting the protective ozone layer of the atmosphere, polluting the air we breathe, and contaminating the sources of our drinking water, just to mention a few of the ways that chemicals are believed to be altering our environment. Many solutions have been proposed and these, at times, seem contradictory. However, before we can objectively consider the issues, we must first understand what chemicals are and how they impact our lives.

The world as we know it is *all* chemicals! The earth, water, and atmosphere of our planet are chemical mixtures. The gaseous atmosphere of Venus and the volcanic surface of the moon are composed of chemicals. The entire universe is one large mixture of chemicals.

Everything on Earth, including humans, is made up of chemicals. Animals, plants, and inanimate objects are constructed of the same basic chemical molecules, consisting primarily of carbon, oxygen, hydrogen, and nitrogen. Animals, plants, and inanimate objects differ in looks and function because of the differing amounts and arrangements of the chemicals composing them.

Elements and Atoms

We use the word *chemical* to refer to both chemical elements and chemical compounds. A chemical element is a specific substance in its simplest form, such as carbon, oxygen, and hydrogen. There are 109 chemical elements, and all the materials constituting our environment are built from these 109 elements, much as the almost 1,000,000 words in the English language are built from just the 26 letters of our alphabet. Each and every one of the 109 elements has its own size, shape, and electronic charge, and is thus uniquely different from the others, just as each letter of the alphabet is different from the others. The smallest amount of an element is called an atom; therefore, when we refer to an atom of carbon, we are talking about the smallest possible amount of carbon.

Some elements (e.g., carbon, oxygen, hydrogen, and nitrogen) are very common and are the major components or "building blocks" of both plants and animals. Hydrogen is the most abundant element in the universe. Other elements, such as bismuth, dysprosium, palladium, and vanadium, are considered trace elements because they are found in very small amounts in our environment.

At room temperature, ten of the elements are found as gases, two are liquids, and the rest are solids. Hydrogen, oxygen, helium, and chlorine are familiar gaseous elements. Bromine and mercury are the two liquids. Solids include calcium, carbon, sodium, zinc, and lead.

Compounds and Molecules

Combinations of chemical elements create chemical compounds, as combinations of letters create words. The smallest measurement of a compound is a molecule.

Each molecule is an exact combination of atoms and has its own specific three-dimensional structure. For example, water is a chemical compound comprised of the elements hydrogen and oxygen. Each water molecule contains two hydrogen atoms and one oxygen atom. A glass of water will contain billions of water molecules, each and every one of them with the same structure, H-O-H or H_2O (Figure 2.1).

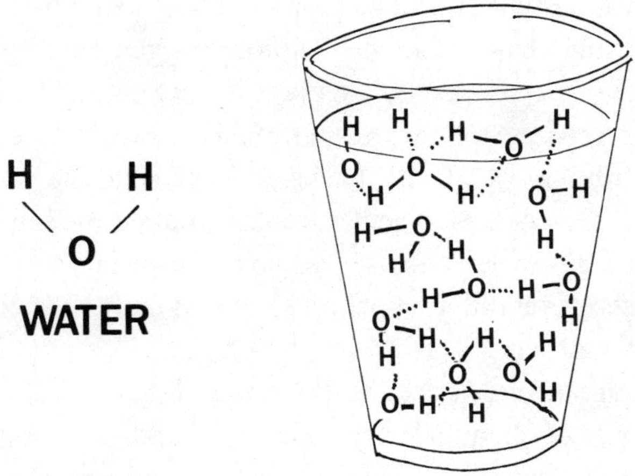

Figure 2.1: Chemical Structure of Water

We can't see a single molecule because it is infinitesimally small. Even the tiniest bacteria is made up of millions of molecules. The human body is made up of trillions of chemical molecules attached together.

Nature's Building Blocks

All things are made from basic "building blocks." For example, the basic materials of a house include concrete, bricks, lumber, nails, plumbing, and windows. The shape of the building and its function will depend on how the basic components are put together. Using the same materials, one could construct a two-story house, a one-floor cottage, a mansion, an apartment building, a shopping center, or a thirty-story office building; the differences result from combining the basic components in differing arrangements and differing amounts.

Almost all of the molecules present in living organisms, be they bacteria, plants, animals, or humans, contain the "building blocks" carbon, oxygen, and hydrogen. Our proteins, carbohydrates, fats, hormones, and DNA are composed mainly of these three basic elements: carbon, oxygen, and hydrogen.

If one could disassemble a plant, a bacteria, and a human, and then separate the various molecules of each organism, one would find most of the basic molecules to be identical. Relatively few of the molecules would be unique to each species. It is similar to when one uses the same fabric and thread, but, by using several different patterns, creates numerous different-looking garments.

The human body contains thousands of different chemicals, and scores of chemicals are naturally found in various foods that make up our everyday diet—milk contains more than ninety different naturally occurring chemicals, whole wheat bread has approximately seventy, coffee contains more than thirty, lettuce has approximately fifty, and roast breast of chicken has about one hundred.

A human body contains thousands of different chemical molecules which all together add up to trillions of molecules. Each molecule occurs in multiples—there are more of some, fewer of others. The names of some of these molecules which are familiar terms to us include estrogen, testosterone, insulin, pepsin, hemoglobin, thyroxin, and adrenalin. While these molecules are made up of a larger number of atoms than is ethanol or formaldehyde, they are still composed mainly of carbon, oxygen, and hydrogen, bonded together differently and thereby making

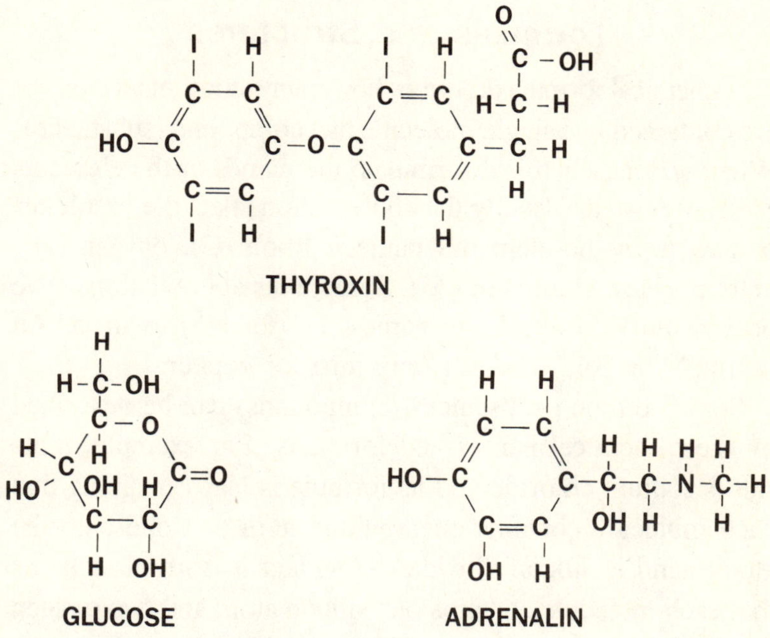

Figure 2.2: Chemical Structures of Thyroxin, Glucose, and Adrenalin

each molecule unique. In addition to carbon, hydrogen, and oxygen, a molecule may also contain other atoms such as the iodine in thyroxin and the nitrogen in adrenalin (Figure 2.2).

Our tissues and organs—heart, muscles, thyroid gland, liver, lungs, skin—appear unalike even though they are composed primarily of the same protein, carbohydrate, and lipid molecules. This is because these numerous basic, similar molecules combine in many different ways to make the various larger and more complex molecules that form our tissues and organs. These tissues and organs, in turn, make up the various body components.

Formulas and Structures

A chemical formula describes how many atoms of an element are contained in a single molecule of a compound (substance). When writing chemical formulas, the names of the elements are abbreviated. Usually the abbreviation takes the first letter or two from the element's name: carbon (C), oxygen (O), helium, (He), aluminum (Al). Sometimes abbreviations have been taken from the Latin names: Fe (ferrum) for iron, Au (aurum) for gold, and Cu (cuprum) for copper.

Some common substances (compounds) can be described by their chemical names and formulas. For example, table salt is sodium chloride and its formula is NaCl meaning that each molecule contains one sodium atom and one chlorine atom; sand is silicon dioxide (SiO_2) and its formula tells us that each molecule contains one silicon atom and two oxygen atoms; chalk is calcium carbonate ($CaCO_3$) with each molecule having one calcium atom, one carbon atom, and three oxygen atoms; and rubbing alcohol is isopropanol (C_3H_7OH) with three carbon atoms, eight hydrogen atoms, and one oxygen atom. Each molecule of a substance is identical; the larger the amount of the substance, the larger the number of molecules.

In order to understand the structure of a molecule, one must look at both the elements contained and their three-dimensional arrangement. Molecules are three-dimensional, but their three-dimensionality is very difficult to represent in a two-dimensional line drawing such as we are using in this presentation. However, please keep in mind that a molecule's structure is actually *three*-dimensional.

Some molecules have exactly the same number of carbon, hydrogen, and oxygen atoms, but the atoms are arranged differently and therefore create different compounds. For example, glucose and fructose are different sugars, but both contain $C_6H_{12}O_6$. They are different because the atoms are arranged differently, as illustrated in Figure 2.3.

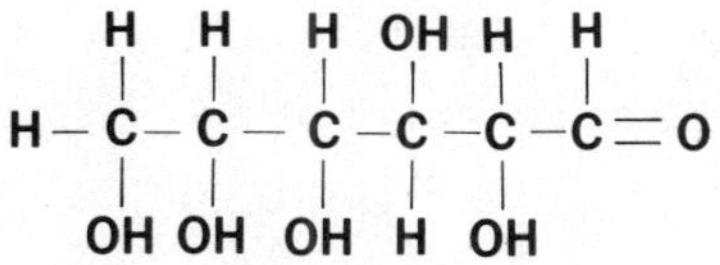
GLUCOSE

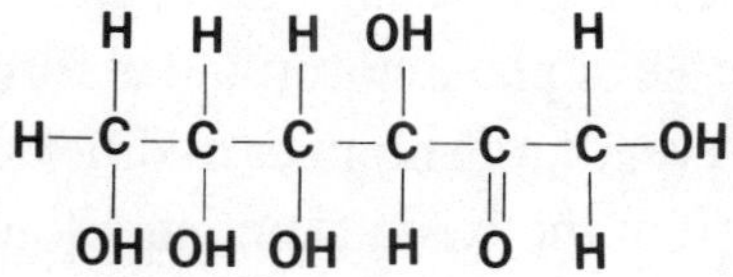
FRUCTOSE

Figure 2.3: Structures of Two Simple Sugars

Acetic acid (vinegar), ethyl alcohol, and formaldehyde are examples of specific molecules, each made up of only carbon, oxygen, and hydrogen in a specific but different combination. Although they contain the same elements, the three molecules have very dissimilar chemical properties. Certainly, by smell alone, we can differentiate between the three substances, although all consist of only carbon, hydrogen, and oxygen. We cannot see the three-dimensionality here, but let's take a look at the chemical structures of the molecules, shown in Figure 2.4.

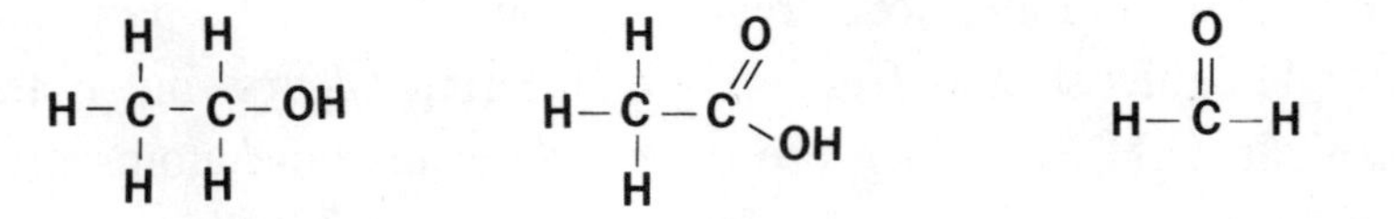

ETHYL ALCOHOL   ACETIC ACID   FORMALDEHYDE

Figure 2.4: Structures of Simple Molecules

The Diverse Uses of Two Familiar Chemicals

The majority of elements have very diverse uses when incorporated into different molecules (substances). Looking at the many different uses of phosphorus and chlorine, we can begin to appreciate the very large number of different molecules that can be formed from the basic elements. Some are hazardous to our health while others are essential for life.

Phosphorus is essential for life. It can be found in every living cell—in high energy molecules, such as adenosine triphosphate, and in the DNA molecule. It is also found in a class of relatively toxic insecticides, the organophosphates, many of which are neurotoxins. Combined with calcium, phosphorus takes on additional uses—monocalcium phosphate is used by bakers as a leavening agent, and dicalcium phosphate is the polishing agent in many toothpastes, to name just two.

Phosphorus, when combined with sodium and oxygen, has been used as a "builder" in laundry detergents. A builder, when added to soap, produces a more efficient cleaning agent. Up to 60 percent of early laundry detergents consisted of sodium tripolyphosphate (STPP) which also acted as a chelating agent to help dissolve metals found in hard water, thus softening the water. STPP was the detergent component blamed for the phosphate pollution of lakes and streams. Because of the environmental concerns related to the phosphates in detergents, nitrilotriacetic acid has been substituted as a builder.

Phosphorus is frequently associated with fertilizers, of which it is a major component. Pure phosphorus, being pyrogenic, can burst into flames spontaneously. However, when combined with carbon, hydrogen, and oxygen, phosphorus becomes an integral component of some of the most effective flame retardant chemicals known.

In contrast to phosphorus which is essential to life and found in every living cell, the element chlorine is not a component of any living cells except those in the stomach lining, and here it combines with hydrogen to form hydrochloric acid which plays an important role in the digestive process. Chlorine combined with sodium forms NaCl, sodium chloride or table salt, a substance we add to our food because it is essential for life.

When oxygen is added to chlorine and sodium in various combinations, compounds are formed which are used in bleach, water purifiers, antiseptics, and disinfectants. When oxygen and carbon are added to chlorine, phosgene, a highly toxic war gas, is formed. Chlorine gas is highly toxic and very irritating to the lungs. When chlorine is added to molecules containing carbon and hydrogen, solvents such as chloroform, carbon tetrachloride, trichloroethylene, and methylene chloride are formed. (Figure 2.5).

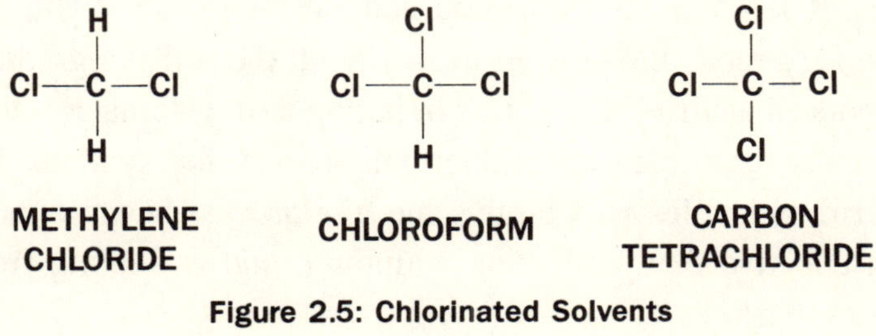

Figure 2.5: Chlorinated Solvents

Natural Vs. Man-Made

Every molecule has a very specific three-dimensional structure that is unique to that molecule. Therefore, ascorbic acid differs from hydrochloric acid and sulfuric acid, as apples, peaches, and pears differ. However, a particular molecule always has the identical chemical structure, regardless of whether it is made by "Mother Nature" in a fruit, or by a man in a laboratory. Man-made chemicals are built using the same elements or "building blocks."

Ascorbic acid (Vitamin C) can be obtained by squeezing citrus fruits and extracting the vitamin from the juice, and for many years, this was the only manner in which it could be obtained. Now, it can also be formed by chemical synthesis—that is, by a chemist combining the carbon, hydrogen, and oxygen atoms in the precise configuration that makes Vitamin C. Synthesized Vitamin C has a chemical structure and biological activity identical to the natural molecules obtained from citrus fruits.

The same is true for the salicylates, the painkilling ingredients in aspirin and oil-of-wintergreen. The original salicylates were obtained from the bark of the willow tree, and the extraction process required the collection of many, many tons of bark. Once the precise chemical structure of the salicylate molecule was determined, chemists learned how to assemble carbon, hydrogen, and oxygen to make the salicylate molecule. This spared the destruction of many millions of willow trees. The chemists even modified the molecule slightly to produce a molecule (acetylsalicylic acid, the active ingredient in modern aspirin) that is less irritating to the stomach. Once chemists developed a commercial process for synthesizing aspirin molecules on a large scale, the process became more cost effective than collecting, shipping, and extracting from willow bark.

Figure 2.6 shows the chemical structures of Vitamin C and aspirin. These molecules are the same whether synthesized or obtained from natural sources.

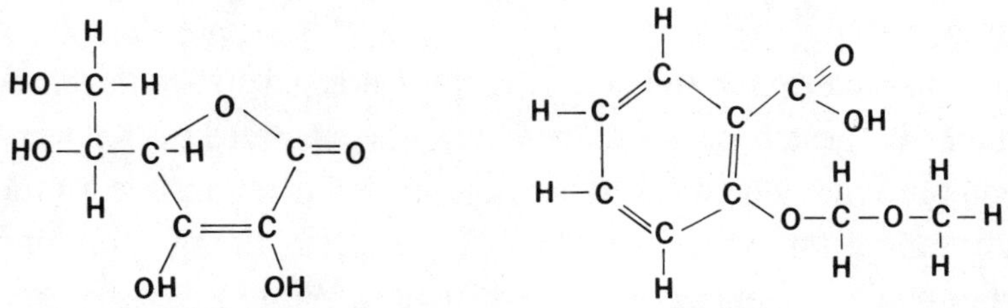

ASCORBIC ACID
(VITAMIN C)

ACETYLSALICYLIC ACID
(ASPIRIN)

Figure 2.6: Chemical Structures for Vitamin C and Aspirin

Small Can Be Great

Many hundreds of different molecules can be made out of just the three elements: carbon, hydrogen, and oxygen. When other elements are added, a still larger number of chemicals become available. If we consider just the three basic elements (carbon, hydrogen, and oxygen), we see that even minor alterations in the arrangement of these atoms into molecules can produce significant effects in their biological activity. Let's look at three simple molecules: (1) methyl alcohol, containing one carbon, can cause blindness if ingested, whereas (2) ethyl alcohol, containing two carbons, is the alcohol found in alcoholic beverages, and (3) acetic acid, also containing two carbons, is the tart material we call vinegar (Figure 2.7).

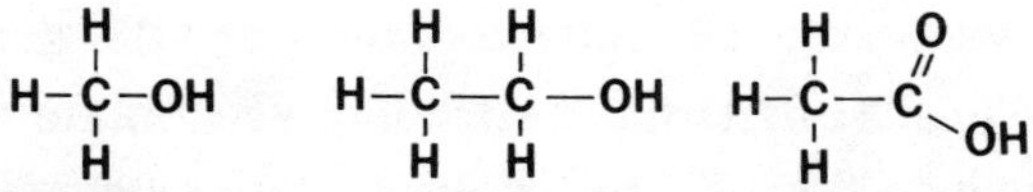

METHYL ALCOHOL

ETHYL ALCOHOL

ACETIC ACID

Figure 2.7: Chemical Structures of Methyl Alcohol, Ethyl Alcohol, and Acetic Acid

Small differences in more complex molecules can make great differences in the molecule's biological activity. For example, testosterone, containing nineteen carbons, is the male sex hormone responsible for the development of the male secondary sex characteristics and is essential for normal sperm development. Progesterone is a female hormone, secreted by the ovary and placenta, which controls ovulation and is reponsible for the thickening and vascularization of the uterus lining occurring after ovulation and prior to menstruation. While the two steroid molecules, testosterone and progesterone, are almost identical in structure, the relatively small difference at just one end makes a very large difference in their biological activity (Figure 2.8).

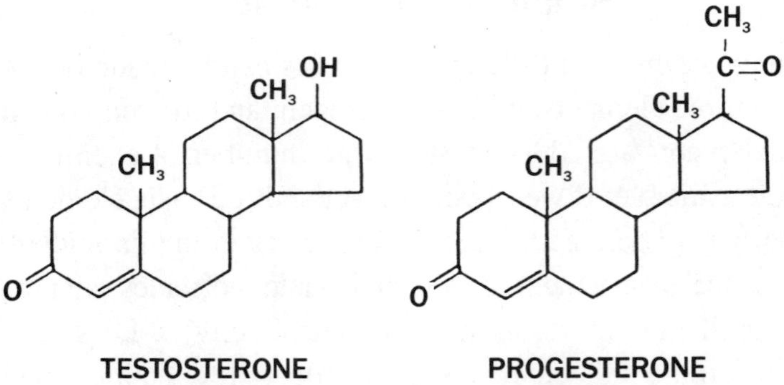

Figure 2.8: Chemical Structures of Testosterone and Progesterone

Other steroid hormones, also containing minor structural changes, have other distinct biological activities, e.g., mineral excretion, water balance, cell metabolism, and other essential functions. We are often told not to eat foods containing cholesterol, yet the ingestion of some cholesterol is essential for normal body function. The body converts cholesterol to testosterone and estrogen, and significant amounts of both hormones are found in both men and women, but in different proportions.

The Diversity of Chemicals

Some chemical molecules (pharmaceuticals) cure illness. Others (nutrients and vitamins) help us maintain our health. Many chemical molecules are designed to have very specialized functions: food additives retard spoilage, dyes make food and clothing more attractive, pesticides selectively kill weeds and insects, and plastic products serve countless useful functions. Some chemicals (toxins) cause harm to the body when it is exposed to significant amounts of them.

The term "chemical" is a very general term used to describe the basic components of all the living and non-living matter which constitutes our universe. Some chemicals we classify as "good" and some we call "bad," but, in their great diversity, they make up the world as we know it.

All matter is composed of chemicals. We are made of chemicals and we are surrounded by chemicals. The interaction of our internal chemicals with external chemicals can sometimes benefit and sometimes harm us. In the following chapter, we will take a closer look at the interaction between external chemicals and the body's internal chemicals. Later on, we will look at how this knowledge is used to determine whether a substance is safe or dangerous to us.

How the Body Detoxifies Chemicals

The mere fact that a body is exposed to potentially toxic chemicals does not necessarily mean that it is damaged by them. The body has a wide range of mechanisms to excrete or detoxify chemicals before they can cause damage to its tissues.

A discussion of how the body rids itself of foreign chemicals is necessary in order for us to appreciate the efficiency of that amazing system. We need to understand how we can remain healthy even though we are constantly exposed to a great variety of external chemicals.

Deoxyribonucleic Acid

The earliest organisms on earth were simple one-cell creatures. The entire animal was contained in a single, tiny, microscopic cell. This single cell, in order to survive in its environment, had to develop the essential biological functions of ingestion of nutrients, digestion, metabolism, excretion, reproduction to perpetuate the species, and detoxification to defend against the natural toxins in its environment.

An organism retains its characteristics through its DNA. DNA is the abbreviation for deoxyribonucleic acid. All the information necessary to perpetuate the species is contained in the DNA molecule. DNA is found within each cell and it contains all the hereditary information that determines the organism's shape, size, means of locomotion, coloration, and its ability to successfully survive in its environment. The DNA within each cell directs all the functions of the cell, and the DNA molecule can build an exact duplicate of itself in order to pass this complex cell-management information from one generation to the next.

Through evolution, the animal kingdom developed thousands of species. With time, one-cell animals evolved into more complex multicellular animals. Yet we need to remember that even the largest and most complex animals begin as a single cell, the fertilized egg, which then divides rapidly to form the growing embryo. As the cells divide and multiply, the DNA, containing the full complement of cell-management information, is passed to each and every new cell within the organism.

In the cells, the DNA is contained in rod-shaped structures called chromosomes, and the information that is transferred from one cell to another, and from one generation to another, is commonly referred to as "the genes." Since even a very complex organism, such as a human, begins as one cell containing all of the genetic information necessary to instruct the cells how to divide, how to differentiate into various organs and tissues, and how to replicate, each and every cell must contain all of that information.

With the emergence of multicellular animals, certain cells within the larger organisms became specialized. At an early stage of embryo development, the genetic material triggers the process of differentiation and the embryo changes from a group of identical generalized cells to an organism containing various specialized cells which form tissues and organs, such as muscle, fat, skin, nerves, bones, kidneys, the liver, and the brain.

Even though each and every cell contains the total

complement of genetic material, only a certain portion of the genetic material is utilized by each type of cell. What triggers cell differentiation is not presently known. Scientists do not know why new liver cells become liver cells, or why only the secretory cells of the pancreas secrete insulin but lung, kidney, and spleen cells cannot (even though these cells also contain those same genes which instruct the pancreas to secrete its hormone.)[1]

The availability and integrity of the information on the DNA is essential to the health and survival of the cell and the organism. If the DNA is damaged, the instructions necessary to maintain the functions within the cell may be altered, resulting in cell death or mutation.

How a Human Body Cell Protects Itself

Within the cell, systems are in place to protect the DNA from exposure to external chemicals, be they natural or man-made, which may damage this essential molecule. These safeguards include the containment of the DNA in a special unit within the cell called the nucleus (which is bound by a difficult-to-penetrate membrane) and the body's detoxification process (which alters the structure of foreign chemicals, making them less hazardous and enabling them to be more readily excreted).

On rare occasions, a few molecules of a reactive chemical get past the protective detoxification enzymes, circulate through the bloodstream, and enter a cell. One or more of these toxic molecules may cross into the nucleus and combine with and damage a segment of the cell's DNA. However, even if this occurs, every cell has a highly efficient DNA repair system which can actually identify damaged segments of the genetic material, excise it, and replace it with a new segment containing the proper hereditary information. This amazing excision-repair system protects us from small amounts of chemicals which have an affinity to bind to the DNA and alter the critical control messages contained in that molecule.

Toxins Cause Damage

Certain toxic chemicals exert their toxicity by interfering with the instructions coming from the DNA to the intracellular machinery which maintains the normal operation of the cell. Normally, the DNA effectively directs all the critical functions of the cell, such as synthesizing essential proteins, lipids, and other cell constituents, and generating the energy to fuel the cell's processes. Cells require energy to secrete hormones, transmit nerve impulses, and work together to move our bodies. Therefore, toxins affecting our DNA can adversely affect critical body functions.

When toxins damage our DNA, they are adversely affecting our genes. Genes, located on the chromosomes, are portions of the DNA which contain specific pieces of information that can be passed on to the next generation, such as eye color, height, and facial characteristics. Interfering with the directions coming from the DNA can cause life-threatening imbalances within the cell, and even cell death. While the loss of one or two cells in a human would not be noticed, the loss of several hundred cells within an organ would lead to serious dysfunction. Also, damaging the DNA in the sperm or eggs could have devastating effects on the species' ability to survive.

Toxins not only affect intracellular DNA. Some toxins have the potential to adversely interact with other components of the cell such as those responsible for energy production, protein synthesis, transport of nutrients, or membrane function. Our cells are vulnerable to both natural and man-made toxins.

The magnitude of cell toxicity depends on several factors, including: the amount of toxin to which the cells were exposed, duration of exposure, the amount of the toxin getting to the organ, the portion of the organ affected, the degree to which the function has been diminished, and whether the effect is reversible. For example, significant exposure to chloroform or carbon tetrachloride can result in severe liver damage. If the exposure concentration is very high and of long duration, a sufficient number of liver cells could be destroyed, causing severe irreversible liver dysfunction. Lower concentrations and/or shorter duration of exposure would produce less toxicity and allow liver regeneration and recovery. Therefore, the level of exposure or "dose" largely determines whether a chemical will cause an adverse effect in a living organism.

How the Body Protects Itself

Healthy individuals are protected by an effective detoxification process which, at low exposure concentrations, is very efficient. At higher exposure concentrations, this process becomes less efficient. Therefore, a chemical's ability to enter the cell and interact with the cell's regulatory machinery and the DNA is dependent upon: (1) the exposure concentration, (2) the duration of exposure, *and* (3) the effectiveness of the individual's detoxification system.

We are continuously exposed to foreign chemicals, both natural and man-made, but this does not mean we are being injured by these chemicals. For *all* chemicals, it is the amount of the chemical that differentiates between harmless and harmful, safe and lethal, therapeutic and poisonous.[2] Thousands of chemicals, both beneficial and harmful, enter the body through the skin, lungs, and digestive system. The very efficient mechanism for processing these chemicals is called "metabolism." When dealing with harmful chemicals, we call the process "detoxification."[3]

The large, highly efficient body organ mostly responsible for collecting and detoxifying foreign chemicals is the liver.[4] All chemicals entering the body through the gastrointestinal tract—whether a nutrient, vitamin, mineral, or pesticide—immediately pass from the intestine to the liver, because the blood from the gastrointestinal tract passes through the liver before it enters the general circulatory system (Figure 3.1). This allows the liver to act as a filter, removing foreign chemicals from the blood before they have a chance to distribute to other parts of the body. Chemicals absorbed through the lungs and skin are also processed by the liver. Once within the body, chemicals are constantly being converted into either useful cellular constituents or excretory products.

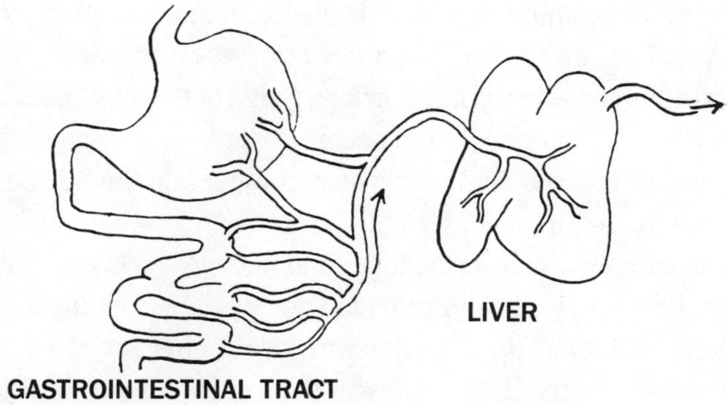

Figure 3.1: Pathway of Ingested Chemicals from GI Tract to Liver

When small amounts of most chemicals arrive in the liver, they are very efficiently removed from the blood and biochemically converted to less toxic and more easily excreted forms, called "metabolites." "Metabolite" is the general term used to describe the biochemical transformation product of any chemical, including the material formed by the liver during the detoxification of a chemical (Figure 3.2).

CHEMICAL

↓

DETOXIFICATION IN LIVER

↓

METABOLITES

Figure 3.2: Metabolite Formation

When large amounts of a toxic chemical enter the liver, the conversion enzymes become overwhelmed and the detoxification process is less efficient. Some of the toxic material enters the general circulation unchanged, and it is transported to various organs, some of which provide another defense against the toxin.

Some organs, such as the lungs and kidneys, have their own detoxifying enzymes and provide another barrier to the toxic materials. Other organs, like the brain, have almost no detoxifying capability. Instead, the brain uses a special membrane (called "the blood-brain barrier") to keep most chemicals out.

Different chemicals have affinities for different organs and tend to accumulate in specific parts of the body. For example, DDT, dioxin, and chlordane concentrate in body fat, iodine in the thyroid, lead in bone, carbon monoxide binds strongly to hemoglobin, and other chemicals concentrate in the liver.

The site at which a chemical accumulates is not necessarily the target site for its toxic action. For example, while more than 90 percent of the body's lead burden is found in the skeleton, lead is specifically toxic to the nervous system. Although lead is the most ubquitous toxic metal in the environment, there is no demonstrated biological need for lead, as there is for other metals such as manganese, magnesium, copper, iron, and selenium.

In contrast to lead which is stored in one tissue but expresses its toxicity in a different tissue, carbon monoxide is both stored and toxic at the same site, the red blood cell. Having a stronger affinity than oxygen for the special site on the hemoglobin molecule where oxygen normally binds for transport to our tissues, the presence of carbon monoxide results in the preferential binding of this poisonous gas, creating a deficiency in available oxygen. When exposed to sufficiently large amounts of carbon monoxide, the resulting oxygen deprivation can cause tissue damage, brain dysfunction, and death.

Very rarely are we exposed to large amounts of chemicals, but even then, exposure is usually for a short time—smoke from a fire, and vapors from paint, gasoline, insecticides, and aerosol products. The liver can deal effectively with these very short exposures to high concentrations. At these times, the detoxification systems in the kidney, lungs, and skin help the liver to safeguard our health. A healthy liver is a complex marvel which we take for granted because we aren't conscious it is working!

How a "Toxic Effect" Occurs

When we continuously absorb large amounts of a toxic chemical, the detoxification capabilities of the liver can be overwhelmed, and this allows significant amounts of the toxicant to be transported via the blood to other organs to interfere with normal organ function. Any adverse effect, causing damage to a body component, is called a "toxic effect."

Perhaps we can better understand how the detoxification pathway becomes saturated and overwhelmed if we consider the following analogy. Prior to shipment from the production farm, chicken eggs are visually examined for cracks and breaks as they are carried down a conveyor belt. Under normal conditions, possibly 5,000 eggs pass the examiner each hour, although he is able to handle as many as 6,000 eggs per hour. Therefore, at 5,000 eggs per hour, the examiner's ability to remove cracked eggs is not stressed. However, if the rate increases to 8,000 per hour, a certain portion of the eggs will go unexamined and the risk of shipping cracked eggs becomes significant. At 10,000 eggs per hour, the risk is even higher, as the system is overwhelmed, and many cracked eggs get through. The same concept holds true for the liver.

How Detoxification Pathways Work

The detoxification enzymes within the liver cells are proteins that are able to very rapidly convert toxic chemicals to less toxic forms. However, when the rate of absorption of a chemical exceeds the maximum rate of detoxification, unchanged molecules get through and are circulated to other parts of the body. It is important to recognize that very few people are ever exposed to saturating concentrations of any chemical.

The two major routes for excretion of chemicals are via the urine and the feces. Two other less utilized routes of excretion are through the skin and lungs. The liver can direct the excretion of certain toxic chemicals and their metabolites directly into the bile. When toxic materials are excreted via the biliary route (passing from liver to bile to intestines to feces), it eliminates the need for those metabolites to enter the bloodstream and circulate through the body to the kidney, lungs, or skin, thus further protecting our tissues and sensitive biochemical systems from interacting with those materials.

In an effort to aid in the elimination of foreign chemicals from the body, the liver modifies these materials in several ways. It can convert many chemicals to a more water soluble form so they can be more readily carried by the aqueous blood to the kidney for expulsion into the urine. The more water soluble a chemical is, the less able it is to accumulate in our tissues, and the more readily it partitions into the blood. For the majority of chemicals, their metabolites formed by the liver are significantly less toxic than the parent chemical itself. Thus the process of detoxification involves the conversion of a chemical to a form which is both less toxic *and* more readily excreted.

In an earlier discussion of chemicals, we noted that all chemicals are, or are made of mixtures of elements such as carbon, hydrogen, and oxygen. In its detoxification process to make chemicals less toxic and more easily excreted, the liver commonly adds combinations of specific elements, called "functional groups," to the chemicals.

One of the most common metabolic pathways is the addition of a functional group called a "hydroxyl," which is just hydrogen-oxygen. Water is composed of hydrogen-oxygen-hydrogen; therefore, since a hydroxyl group is comprised of hydrogen-oxygen, we can correctly describe water as a hydroxyl combined with a hydrogen (HO + H = HOH). The addition of the hydroxyl makes many drugs, food ingredients, pesticides, and other types of chemicals more compatible with blood, and thus more readily and efficiently transported to the kidneys. This "hydroxylation" of a chemical usually converts it to a safer structure which has significantly less toxicity.[5] Typical hydroxylations are shown in Figure 3.3

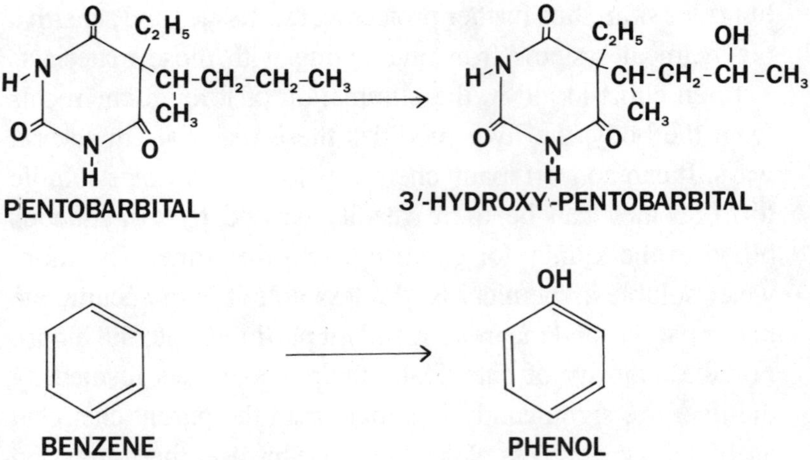

Figure 3.3: Detoxification by Hydroxylation

In addition to metabolic hydroxylation, chemicals are biotransformed in other ways. For example, "oxidative de-alkylation" actually splits off part of a chemical molecule, thus forming two smaller molecules which in most cases are easier to excrete (Figure 3.4).

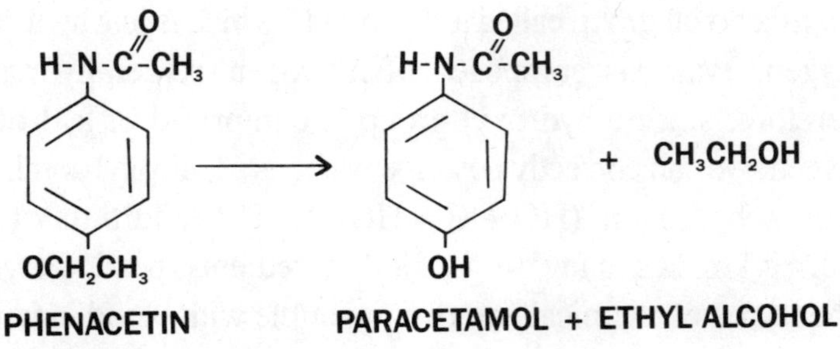

Figure 3.4: Detoxification by Dealkylation

For chemicals containing a primary amino functional group (NH_2), "deamination" can occur; deamination is the removal of the amine group from the molelcule, resulting in the formation of a non-nitrogen-containing metabolite, plus ammonia (Figure 3.5).

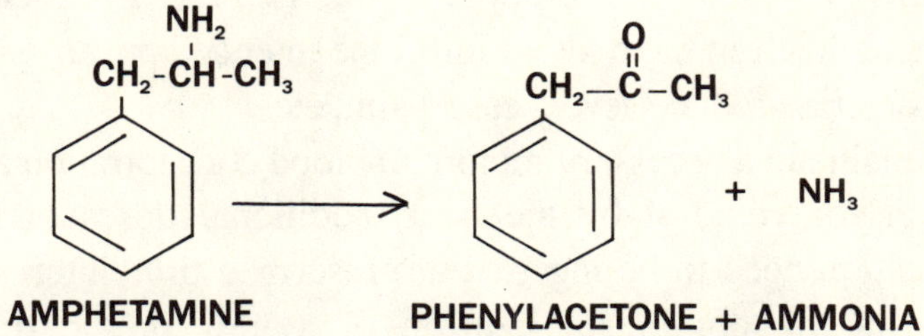

Figure 3.5: Detoxification by Deamination

In addition to hydroxylation, dealkylation, and deamination, there are many other metabolic biotransformation reactions available for use by the liver. Thus, the liver has the amazing ability to act as a complex chemical processing facility, but works with greater speed, efficiency, and accuracy than would be possible for any man-made laboratory.

While the magnificent detoxification process just described provides us with a highly effective level of protection against the physical insults caused by foreign chemicals, it has a few other effects which we should also be aware of.

How Medications Are Processed

When a person takes a prescribed medication for an illness, he must repeat the dosage every few hours over a period of days or weeks. When these foreign molecules, which we call "medications" or "drugs," are absorbed from the gastrointestinal system and arrive at the liver, they are recognized as foreign molecules not belonging to the body, and the liver begins to modify or "detoxify" them. Therefore, some amount of the medication is immediately removed from the blood by the detoxification process.

As the remaining medication molecules circulate through the body, they pass through the liver, again and again. With each pass, more medication molecules are inactivated and modified for ease of excretion. Soon, there is insufficient medication left in the body to fulfill the intended purpose—to kill bacteria, reduce fever, ease pain, etc.

To maintain a necessary minimum blood concentration and counter the removal by the liver, additional doses of the medication need to be ingested at prescribed time intervals. Prior to the commercialization of a new medication, pharmaceutical companies conduct carefully controlled clinical tests to determine the rate at which the new medication is removed from the blood by the liver. This information is used to define the optimum schedule for doses of medication prescribed by a physician or indicated on the product label.

Every living plant and animal has some chemical biotransformation capability. Even bacteria can detoxify medications. Medication is prescribed in a dose large enough to overload the bacteria's detoxification system so that some medication molecules remain available to inhibit bacterial growth and survival.

Some strains of bacteria develop enhanced detoxification systems. When an infection doesn't respond to the administered medication, it is probably because the bacteria has become resistant to that particular medication. One common form of bacterial resistance is the development of a more active detoxification system which allows the bacteria to destroy the drug as rapidly as it enters the bacterial cell.

Understanding the importance of chemical transformation reactions has been of great value in the development of new medications. Certain medications are active when administered by injection but are inactive orally. Yet, oral administration is certainly the preferred route, since it is safer and easier.

Our intestines contain many types of bacteria and some are able to rapidly metabolize certain medications when they are given orally, before the medication has a chance to be absorbed into the body. Recognizing this problem for certain medications, pharmaceutical companies have designed "pro-drugs." Additional functional groups are added to the medication (drug) molecule making the medication resistant to inactivation by the gut bacteria. Once absorbed, the pro-drug is cleaved by the liver, releasing the active medication for transport throughout the body.

The concept of "pro-drugs" came about as a result of an accidental discovery. Sulfur-containing chemicals were synthesized early in this century by German chemists for use as dyes. In 1935, the medical researcher Domagk found that one of these dyes, Prontosil, had antibacterial activity when he used the dye on mice infected with *Streptococci*. Subsequently, it was discovered that the active ingredient was not Prontosil itself but a metabolite, sulfanilamide.[6] The inactive compound is converted to an active chemotherapeutic metabolite as a result of the conversion by the liver, as shown in figure 3.6.

Actually, the first pro-drug was methenamine, introduced in 1899 as a urinary antiseptic. When taken before a meal (when there are essentially no stomach acids), it passes through the gastrointestinal tract, remains unchanged in the blood, and is excreted into the urine. When it encounters the acidic urine, the molecule splits in two, liberating formaldehyde which effectively kills infectious bacteria in the urinary tract. Surprisingly, methenamine is still used to treat urinary tract infections in many countries.[7]

Aspirin is a pro-drug. When the active chemical in willow bark was first identified as a form of salicylic acid, chemists prepared the sodium salt of the acid for oral administration.

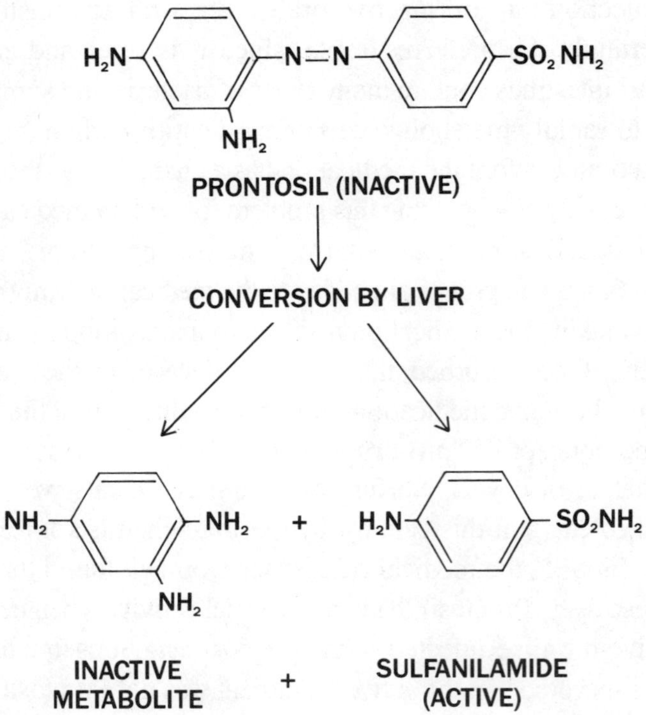

Figure 3.6: Liver Conversion of Prontosil

While sodium salicylate had the analgesic properties of willow bark, it caused severe stomach irritation in many people. For this reason, an acetyl functional group was added to salicylic acid, forming acetylsalicylic acid, the active chemical in present day aspirin, a pro-drug. In the stomach, the acetyl protects from irritation but once absorbed, the body quickly removes the acetyl group, releasing the active drug, salicylic acid (Figure 3.7).

How Biologically Inactive Chemicals Are Bioactivated to Toxins

The same liver enzymes that protect us by detoxifying foreign chemicals entering our bodies can also inadvertantly activate

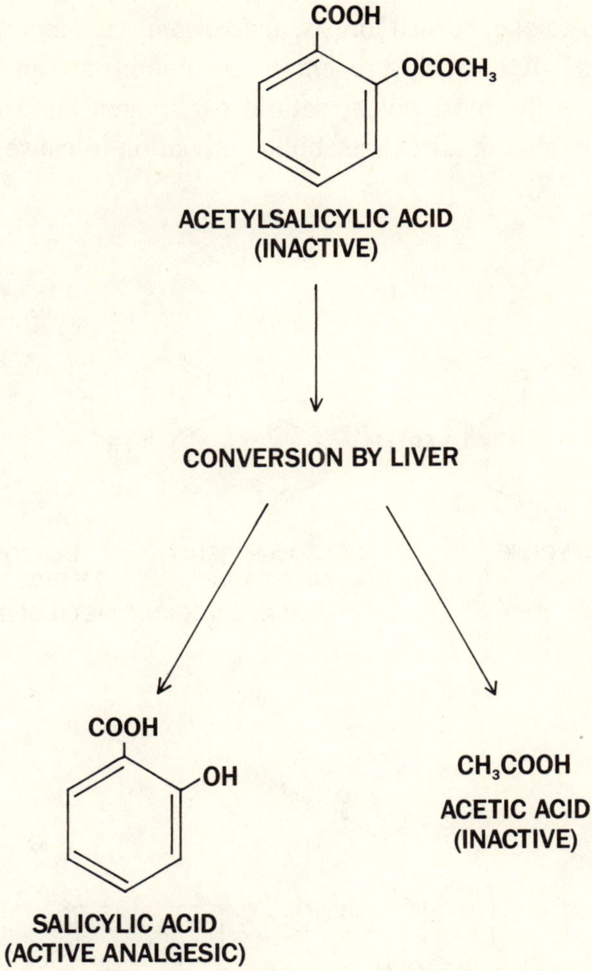

Figure 3.7: Conversion of Acetylsalicylic Acid to Active Analgesic (Salicylic Acid)

certain chemicals to a toxic form. For example, benzo(a)pyrene, classified as a chemical carcinogen, is not itself carcinogenic, but becomes carcinogenic after bioactivation to a reactive carcinogenic metabolite which causes cancer in essentially every animal species.[8] Benzo(a)pyrene forms as a result of burning hydrocarbons, and occurs in grilled meats, auto emissions, tobacco smoke, burned forests, and volcanic ash, among other sources.[9] Aflatoxin, present in peanut butter and many grains, is the most potent natural carcinogen known,[10] and aflatoxin also requires metabolic activiation to cause cancer (Figure 3.8).

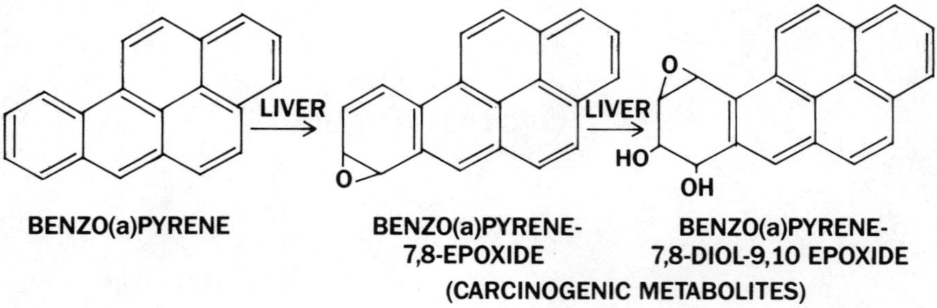

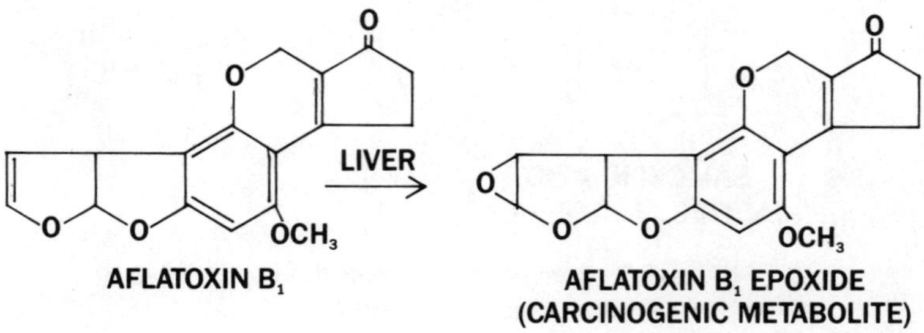

Figure 3.8: Metabolic Bioactivation of Benzopyrene and Aflatoxin

Footnotes

1. When scientists learn how to turn on and off segments of the DNA, they will be able to more efficiently control diseases such as cancer, AIDS, and some genetic disorders.
2. First stated by Paracelsus (1493-1541), now a basic rule of toxicology. This rule is stated in all modern toxicology texts.
3. G. Sipes and A. J. Gandolfi, "Biotransformation of Toxicants," *Casarett and Doull's Toxicology*, 3rd Ed. (New York: MacMillan Publishing Co., 1986), p. 64.
4. *Ibid.*, p. 65.
5. For a discussion of detoxification reactions, such as hydroxylation, dealkylation, deamination, and other biotransformation pathways, please refer to drug metabolism or toxicology texts, such as: (a) A. Goldstein and others, *Principles of Drug Action* (New York: Harper and Row Publishers, 1969); (b) B. N. LaDu and others, *Fundamentals of Drug Metabolism and Disposition* (Baltimore: The Williams and Wilkins Co., 1971); (c) B. Testa and P. Jenner, *Drug Metabolism: Chemical and Biochemical Aspects* (New York: Marcel Dekker, Inc., 1976).
6. J. Trefouel and others, "Activite du p. aminophenyl sulfamide sur les infections streptococciques experimentales de la souris et du lapin," Compt. Rend. *120*, 756, 1935.
7. A. Albert, *Selective Toxicity* 7th Ed. (New York: Chapman and Hall, 1985), p. 98.
8. P. W. Jones and R. I. Freudenthal, Editors, *Polynuclear Aromatic Hydrocarbons* (New York: Raven Press, 1978).
9. *Ibid.*
10. National Research Council, *Diet, Nutrition and Cancer*, (Washington, DC: National Academy Press, 1982), p. 237.

Risky Living

Voluntary and Involuntary Risks

We are continually faced with two types of risks: voluntary and involuntary. Voluntary risks arise out of activities we engage in by choice and in which we feel we have control. It seems that we willingly accept significantly higher risks from voluntary activities– driving, skiing, swimming, smoking, excessive drinking of alcoholic beverages–than from situations over which we believe we have little or no control, such as exposure to pollution of various types, living near a nuclear power plant, ingestion of pesticide residues, etc. However, the ability to control does not really diminish the actual risk of a situation or product, but merely provides a false sense of security.

We routinely endanger ourselves and our children by taking risks with a high probability of accidental injury or death. For example, traffic accidents are the leading cause of death for people between the ages of 5 and 35,[1] yet most of us travel in our automobiles every day. In 1987, the National Center for Statistical Analysis reported 46,386 fatalities from automobile accidents. This is comparable to a fully loaded passenger plane crashing with no survivors every day! Nevertheless, we seem much more fearful of flying than driving.

The illusion of control provides false comfort and often makes risky situations even more dangerous. For example, most of us believe ourselves to be safe drivers, able to handle our cars to their limits. However, out of every 4,000 people who use an automobile, 1 dies in a vehicular accident. Our government's standards state that an item, whether it be a cosmetic, food additive, or pesticide, has an acceptable risk and is considered "safe" if lifetime use increases the number of deaths by no more than 1 in 1,000,000 people. Since motor vehicles cause an increase of about 240 deaths per 1,000,000 people, we should acknowledge, if we measure by the same standards, that they do not qualify as being "safe" and are therefore hazardous to our health. Clearly, in this instance we are exercising our prerogative to decide that the benefits of using this product outweigh the risks involved, and those risks then become voluntary and acceptable to us.

The United States Environmental Protection Agency (EPA) judges a pesticide to be "of acceptable risk" if daily, lifetime exposure to the chemical results in no more than 1 extra cancer per 1,000,000 population. Other risks that increase the chance of death by 1 in 1,000,000[2] are shown in Table 1.

Table 1: Some Relative Risks

Risk:	Death Results From:
Staying 1 hour in a coal mine	Black lung disease
Riding 10 minutes on a bicycle	Accident
Traveling 300 miles by car	Accident
Traveling 1000 miles by plane	Accident
Taking 1 chest X-ray	Cancer from radiation
Eating 40 tablespoons peanut butter	Cancer from aflatoxin
Drinking Miami water for 1 year	Cancer from chloroform

The most widespread environmental carcinogen is ultraviolet light from the sun. Most skin cancers occur on the face, ears, hands, and other body parts that aren't sufficiently protected from sunlight. Skin cancer is more common among people who work outdoors and in tropical regions. Skin cancer itself is rarely fatal, but it can spread to other parts of the body. Even though the cause-effect relationship between overexposure to the sun and skin cancer has been well documented in scientific literature and also has been well publicized, we continue to willingly take the risk.

Table 2 compares people's perception of risk from nuclear power, smoking, and other sources, with the actual risk.[3]

Table 2: Actual Risk vs People's Perception of It

Actual Risk	Activity and Resulting Deaths Per Year	Rating by League of Women Voters	Rating by College Students	Rating by Business/ Professions
1	Smoking (150,000)	4	3	4
2	Alcoholic Bev (100,000)	6	7	5
3	Motor Vehicles (50,000)	2	5	3
4	Handguns (17,000)	3	2	1
5	Elec Power Gen (14,000)	18	1	19
6	Motorcycles (3,000)	5	6	2
7	Swimming (3,000)	19	30	17
8	Surgery (2,800)	10	11	9
9	X-rays (2,300)	22	17	24
19	Commerc Aviation (130)	17	16	18
20	Nuclear Power (100)	1	1	8
28	Pesticides (uncertain)	9	4	15

Many people perceive nuclear power to be of high risk, in line with surgery, handguns, hanggliding, and automobile racing. This misperception arises from the potential for catastrophic accident, the fear of radiation hazard, and the media's sensationalization of events associated with nuclear-powered generators, such as inconsequential leaks, the accident at Three Mile Island, and the tragedy at Chernobyl.

Risks and Benefits From the Use of Nuclear Power

If we put our emotions and fears aside, we can objectively examine the actual risk of nuclear-powered energy generation compared to conventional power plants. While there are many processes that could malfunction in a nuclear power plant, there are also many highly effective safeguards installed during the construction of the facility that make the occurrence of an accident very improbable, thus producing a relatively low risk situation.

On March 28, 1979, a reactor at the Three Mile Island nuclear power plant began sending ambiguous signals to the control room operators, suggesting that the reactor's core pressure was too high and that in response, too much cooling water was being channeled to the core to keep it from overheating. Misreading the hundreds of blinking lights, the operators overrode the emergency cooling system and turned off the cooling water, believing that the steam pressure relief valve for the core was shut. In fact, the operators had not stopped the flow of water, and the vent valve was actually open. Therefore, steam from the overheated reactor coolant water was released into the air for about 2 hours.[4] The bellowing steam emanating from the tall stacks was shown on national television with insinuations that it was highly contaminated with radioactivity. Some commentators described the steam as "radioactive leakage." Shortly, the valve was closed and the cooling system was in normal operating mode.

Immediately after the accident, the average radiation exposure to the public in the vicinity of Three Mile Island was calculated to be 1.4 millirem, an amount about 3 times *lower* than what a passenger routinely receives on 1 coast-to-coast flight (above the clouds which filter much of the radiation from the sun to the earth).[5] Even after government scientists reported the very low level of radiation released, environmental activists continued to allege that the surrounding communities were at great risk from radiation exposure.

Exhaustive investigations were conducted in 1979 by 6 independent groups, including a Presidential Commission and Congressional committees. The common conclusion reached was that the accident occurred not because of a faulty reactor, but because of inadequate operator training and the poor design of the control room and instruments.[6,7,8,9] Had proper signals been delivered to adequately trained operators, the steam release would not have occurrred. As a result, the 138 nuclear power plants in operation or under construction implemented extensive control room modifications and instrumentation enhancement, and immediately provided extensive operator training on the new control systems. In addition, backup systems are now in place so the confusion and inappropriate systems override can no longer occur.

Because reactors built in the United States are multiple barrier containment facilities (have several layers of steel and concrete between the core and the outside environment–Figure 4.1), there is very little likelihood of an accidental release of radioactive material. These plants are also specifically designed to withstand damage from external forces, such as jet airliner crashes, tornadoes, and earthquakes. The earthquake standards are several times more stringent for nuclear plants than for hospitals, fossil fuel plants, and other structures. When in operation, the nuclear power plant produces very little waste (which must be contained in a carefully regulated manner) and no effluent except steam.

The process of generating energy from nuclear fuel, called "fission," occurs within the nuclear fuel in the core of the reactor. When a tiny particle called a neutron is absorbed by an atom,

Figure 4.1: Nuclear Reactors
Barriers against release of radioactive materials from U.S. reactors

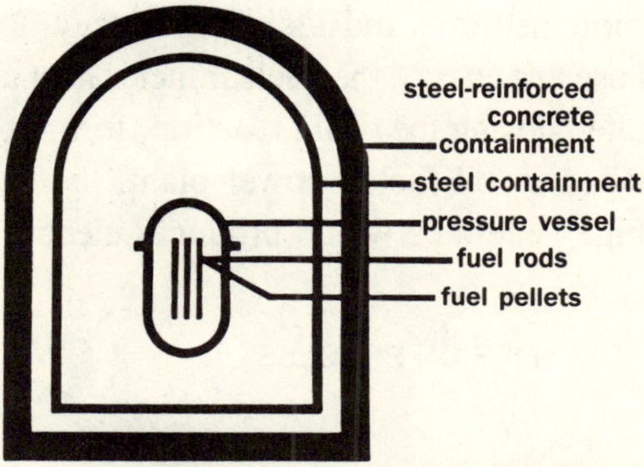

Source: U.S. Council for Energy Awareness

All commercial nuclear power plants now operating in the U.S. use a series of multiple physical barriers to prevent the release of large quantities of radioactive material during normal and accident conditions.

Although details may differ among plants, most will have these elements to contain radioactive products of the fission reaction:

- The nuclear fuel itself, fabricated into dense ceramic pellets inside which most fission products remain bound.
- The fuel rods of zirconium alloy, about 12 feet long and half an inch in diameter, into which the pellets are stacked and sealed.
- The steel pressure vessel—up to 75 feet high, 20 feet in diameter, with walls nine inches thick—containing the several thousand rods making up the fuel core.
- The primary containment, an essentially leak-tight shell of steel plate which encloses the nuclear steam supply system.
- The massive containment building, a steel-reinforced concrete structure with walls three or more feet thick.

To reach the environment, radioactive fission products would have to penetrate each of these barriers in succession. These barriers also are designed to withstand severe external events as well as forces generated inside the system.

the atom splits into two smaller atoms and two neutrons. This fission process releases energy in the form of heat. The two neutrons just formed go on to split two additional atoms, producing more neutrons and additional energy. To efficiently capture and use this energy, the nuclear fuel is kept under water, which is heated during the chain reaction, forming steam. As in a coal-fueled or oil-fueled power plant, the heated steam turns a turbine generator which produces electricity (Figure 4.2).

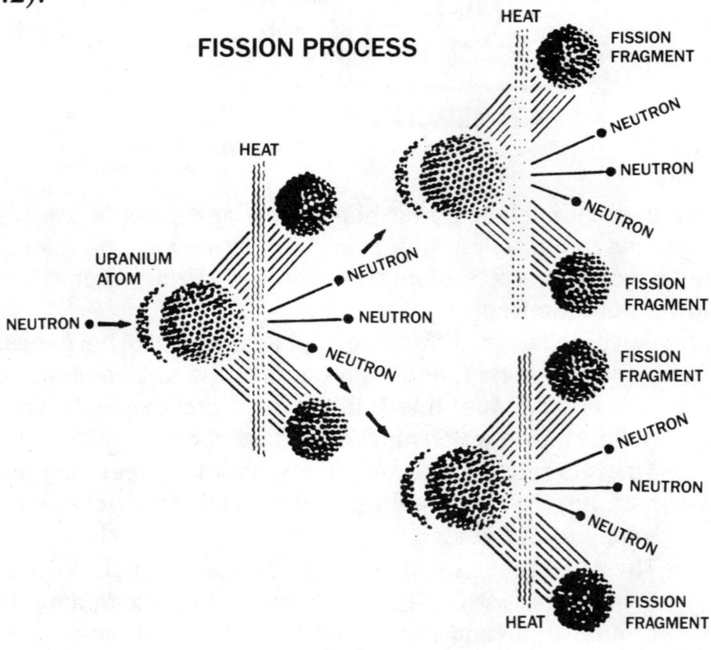

Figure 4.2: Nuclear Fission

The smokestack effluent from the nuclear power plant is steam. In contrast, stack emissions from coal- or oil-fueled power plants contain carbon dioxide, sulfur oxides, nitrogenous oxides, and carcinogenic polycylic aromatic hydrocarbons. In addition, while a relatively small tube of nuclear fuel can generate heat for years, coal-burning results in tons of ash which must be disposed of. The ash contains toxic heavy metals and carcinogenic hydrocarbons. We should question our legislators as to (1) where we can safely dispose of these tons of ash, and (2) what its impact on our environment will be.

Instead of adding nuclear power stations as our energy needs continue to grow, we are building facilities powered by fossil fuels. More than 100 nuclear power plants have been cancelled in the United States, some during construction and others at the planning stage. It has been estimated that the public utility companies have spent over 30 billion dollars on design, permit application, and construction of facilities that were never completed. We, through taxes and our monthly utility payments, are paying for these political decisions.

In South Carolina, Duke Power Company stopped construction of their Cherokee Plant after spending 700 million dollars. Consumers Power Company in Michigan began construction of their Midland Nuclear Plant in 1969. By 1986, 4.1 billion dollars were spent but the facility was still incomplete. It is now being converted to a facility that will burn conventional fuel. The Tennessee Valley Authority spent over 2 billion dollars before abandoning their incompleted nuclear power plants.

When oil prices rose sharply in the late 1970's and early 1980's, several utility companies began reexamining this alternative energy source. Unfortunately, the accident at Chernobyl in the Soviet Union raised anew the public's concern over the safety of nuclear energy.

The Chernobyl accident, the worst nuclear plant accident in the history of civilian nuclear energy, occurred in April, 1986. This accident was the result of two operators conducting an unauthorized experiment with the Soviet nuclear reactor. The thirty-one deaths resulting from the Chernobyl nuclear reactor accident equalled the total number of accidental radiation deaths occurring in the entire world during the previous forty-two years.[10] The devastating and far-reaching results of this accident have caused the American public additional concern regarding nuclear power.

The differences between a United States reactor and a Soviet reactor are significant. The Soviet accident could not have happened in a U.S. reactor. U.S. reactors are housed in facilities that are designed for containment. These sealed concrete and steel structures would retain most of the generated radioactive gases. In contrast, the Soviet reactor at Chernobyl was housed in a structure not designed for containment. Figure 4.3 compares the Soviet and U.S. reactors. Most Soviet reactors, including the one at Chernobyl, are of a relatively primitive design which exists only in the Soviet Union; these reactors do not have multiple barrier containment or effective backup control systems.

We must realize and accept that we live on a radioactive planet. Radiation is all around us, and is a constant part of our environment. Naturally occurring radioactive materials are in our air, food, soil, and building materials. We receive radiation exposure from the sun and from X-rays. The average American receives approximately 150-200 millirems of radiation each year from natural and man-made sources. A nuclear power plant contributes about 0.01 millirems to the environment per year.[11] Table 3 compares the relative strength of the radioactive sources to which we are exposed. Even the water discharged from nuclear power plants contains less radiation than many commonly used liquids, as shown in Table 4.

In contrast to nuclear power plants which emit only steam into our air, coal-powered and oil-powered plants emit a variety of chemical carcinogens into our air 24 hours a day, 7 days a week.[12] This increases the risk of lung cancer, and the

Figure 4.3: Comparison of Soviet and U.S. Reactors

U.S. REACTORS

Boiling water reactor (BWR)

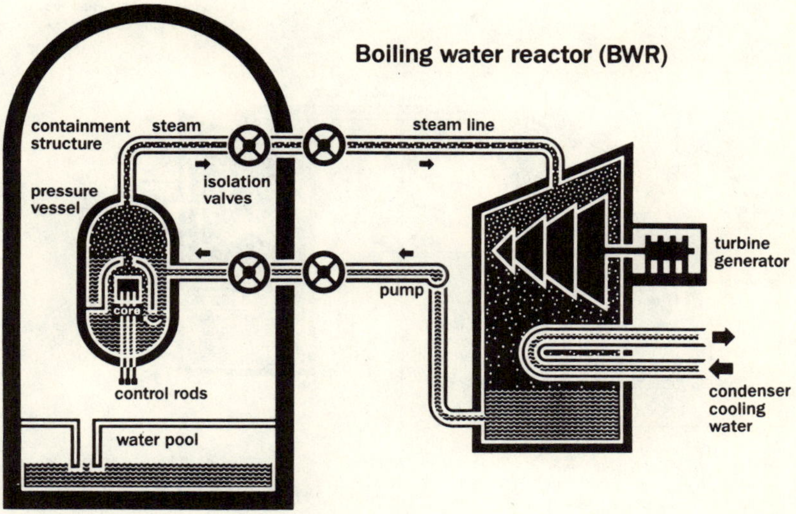

Pressurized water reactor (PWR)

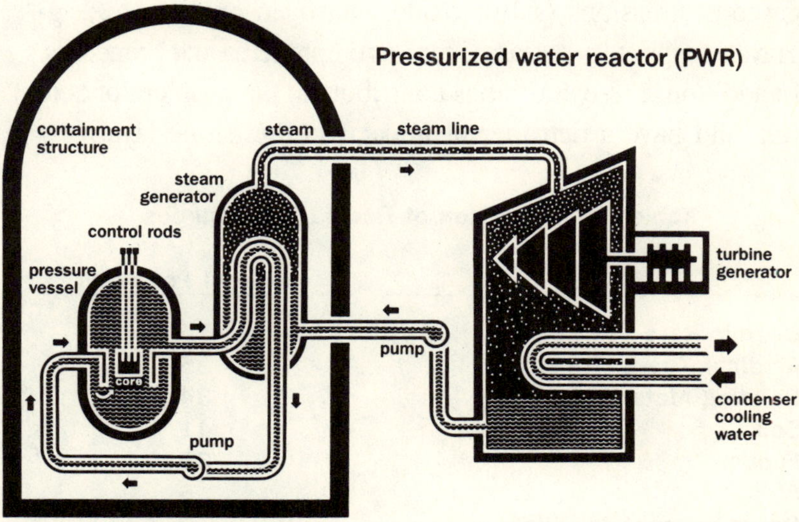

Source: U.S. Council for Energy Awareness

SOVIET REACTOR

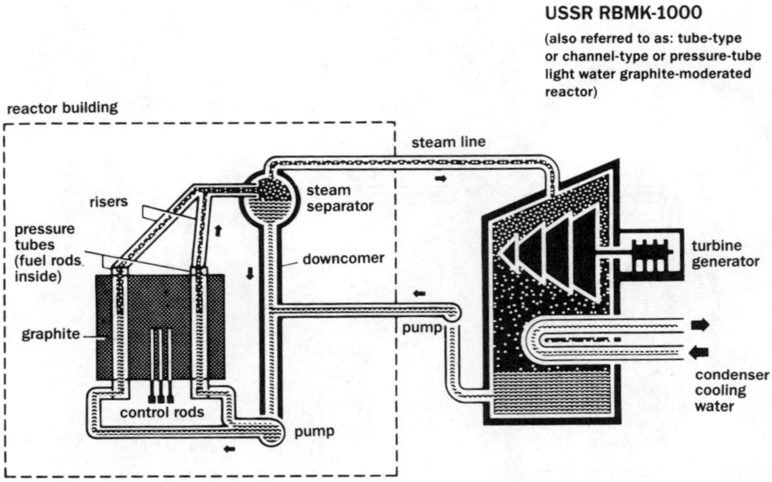

Source: U.S. Council for Energy Awareness

gaseous emissions (sulfur oxides, nitrogen oxides, etc.) aggravate respiratory diseases, such as emphysema and bronchitis. In addition, these emissions contribute to the problem of acid rain and have a detrimental impact on the ozone layer.

Table 3: Comparison of Radioactive Sources

Source	Millirem Per Year
Cosmic Rays	35
Medical X-rays; Therapy	35
Building Materials	34
Soil	11
Foods	25
Air	5
Coast-to-coast Jet Flight	2
Nuclear Power Plant	0.01

Table 4: Comparison of Radioactivity Levels in Commonly Used Liquids

Liquid	Picocuries Per Liter*
Salad oil	4900
Milk	1400
Whiskey	1200
Ocean Water	350
River Water	10-100
Domestic tap water	20
Nuclear power plant discharge	1-10

*A picocurie is a trillionth of a curie.
A liter is approximately a quart.

Acid Rain—Its Causes and Effects

The phenomenon called "acid rain" receives considerable visibility in the media. We are told that acid rain is responsible for a variety of illnesses and causes considerable damage to our environment. While offering no solutions, some commentators suggest that acid rain is a relatively new problem which can, with sufficient interest and effort, be readily reversed. Unfortunately, this is not the case. Although most experts acknowledge that the industrialized world is the cause of this phenomenon, the problem isn't a simple one to solve. In order to participate in the solution of this international environmental issue, we need to understand why acid rain occurs, how it is caused, and how it is able to adversely impact both human and environmental health. We must explore what we as individuals can do to help combat this serious problem.

Contrary to the currently popular notion that acid rain is a new environmental problem, the acidification of our environment has been occurring for a long time. Over 100 years ago, a British government scientist named Robert Angus Smith wrote a book, entitled *Air and Rain, the Beginnings of Chemical*

Climatology, in which he identified as critical issues, "Air of Impure Places," and "Bad Air and the Sensations." Smith, in the 1870's, actually used the words "acid rain" to describe the polluted air in London, and it is believed that he was the first to identify the phenomenon of acid rain.[13]

Although the problem has existed for many years, the increasing levels of atmospheric pollution accompanying the growth and prosperity of our major cities, coupled with the increased number of automobiles, has accelerated the occurrence of acid rain which only recently has been fully appreciated as a serious international environmental problem.

The acidity of a material is measured in units called "pH." A chemical with a pH of 7.0 is neutral (neither acidic nor basic). Chemicals with a pH above 7.0 are basic or caustic, whereas below pH 7.0 they are acidic. The further from 7.0, the stronger the acid or base. Pure water would have a pH of 7.0; a very strong solution of sulfuric acid would have a pH of about 1.5.[14]

When we burn fossil fuels such as coal, oil, natural gas, and wood, some of the resulting pollutants combine with oxygen in the air to form acidic chemicals in our atmosphere.[15] Sulfur and nitrogen are two chemical elements commonly found in fossil fuels such as coal and oil. As a fossil fuel is burned in the generation of energy, sulfur and nitrogen are released as gas, frequently in combination with oxygen.

When entering the atmosphere, sulfurous oxide (SO) combines with an additional oxygen and forms SO_2, sulfur dioxide. Sulfur dioxide is often converted to sulfuric acid (H_2SO_4). With strongly acidic sulfuric acid in the air, atmospheric moisture becomes acidic, and falls as acid rain.[16] The two major sources of sulfur emissions are (1) the large utility companies which burn coal and oil, and (2) diesel engines. Since most of the sulfur is removed from unleaded gasoline during the refining process, oil refineries are also a significant source of sulfur emissions. Further, the installation of catalytic converters in 1975 to automobile exhausts increased the exhaust emissions of sulfuric acid by almost 50 times the amount from cars without catalytic converters. The unit converts the sulfur in gasoline to sulfur trioxide, which reacts with atmospheric water to form sulfuric acid.[17]

Nitrogen entering the atmosphere combines with oxygen to form nitrous oxide (NO), nitrogen dioxide (NO_2), and finally nitric acid (HNO_3). Although catalytic converters on automobiles remove most of the nitrous oxide from exhaust emissions, significant amounts enter the atmosphere from home furnaces and from furnaces used to heat shopping centers, factories, and office buildings.

While in the air, these acidic sulfur and nitrogen oxides also combine with tiny airborne pollutants which exist as small particles (like a fine dust). When these acidic chemicals fall back to earth as either tiny acidic droplets or particles, they are referred to as "acid rain." This cycle is illustrated in Figure 4.4.

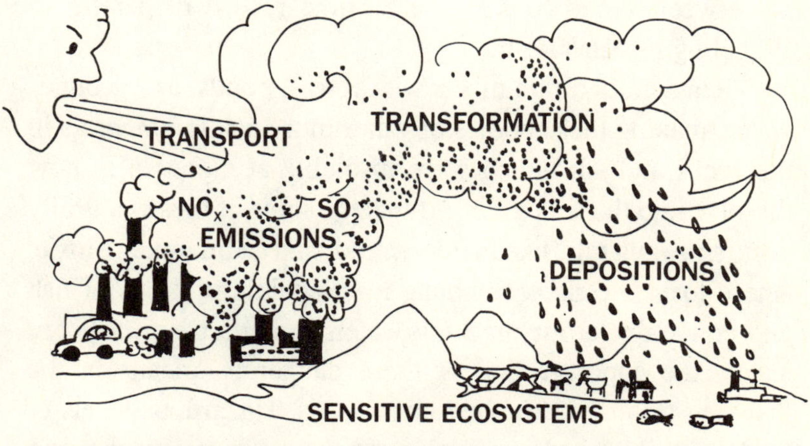

Acid rain results when sulfur dioxide and nitrogen oxides released into the atmosphere are converted into sulfuric acid and nitric acid.

Figure 4.4: How Acid Rain Occurs

Although we are now more aware of the presence of acid rain and are beginning to understand how it occurs, there is yet to be established a direct link between the presence of these airborne chemicals and an adverse effect on human health.[18] Government studies, industry studies, and assessments by specially convened expert panels all suggest a possible association between acid rain and adverse human health effects, but no group has yet provided compelling evidence.[19]

Two specific aspects are receiving special attention. Preliminary evidence suggests that people with respiratory disease, such as asthma and emphysema, are adversely affected when breathing air containing acidic air pollutants. While this is certainly of concern, a second effect of acid rain can have a substantial impact on major segments of our population. Most of us might guess that rainwater would have a neutral pH of 7.0, but actually it has a pH of about 5.6, due to its interaction with carbon dioxide in the air. In an acidified atmosphere, especially in the northeastern United States, rainwater can be significantly more acidic. It is frequently found to have a pH of between 4 and 5, with the all-time record of pH 1.5 in Wheeling, West Virginia.[20]

When acidified rain mixes with rivers, ponds, and groundwater, making these bodies of water more acidic, minerals in the rocks and silt which were insoluble at higher pHs now dissolve into the water. Unfortunately, some of these naturally occurring minerals, like mercury, lead, and cadmium, are toxic. These toxic metals accumulate in the edible portions of fish and contaminate our drinking water. In certain northeastern states, the concentration of these dangerous metals in the drinking water has reached toxic levels. Hazardous levels of lead, cadmium, and mercury have been measured in trout and bass and in drinking water, but there have been no reported cases of human poisonings or ill health directly attributable to the contaminated water or fish. This could be because many physicians are unaware of the hazard and are unfamiliar with the symptoms associated with metal poisoning.

Through its ability to enhance the solubility of minerals in stone, acid rain is responsible for the increased rate of decay of historic structures worldwide. It penetrates buildings and reacts with all forms of artwork, accelerating deterioration, catalyzing the crumbling of old, historic limestone structures, and eating away at brass, copper, and iron materials. Unfortunately, as we develop a sophisticated future, we are accelerating the destruction of our past—an irreversible process.

Even though acid rain contributes to the corrosion of automobiles, may increase lung disease especially in the aged population, and is clearly causing damage to our environment, there is at least one positive result of acid rain. It has recently been shown that increasing the acidity of certain soils makes important micronutrients available to crops. A recent report from California showed that when the acidity of certain soils was increased slightly, micronutrient availability improved dramatically for critical nutrients such as iron, copper, magnesium, and zinc.[21] When bioconcentrated in fish or in our tissues, these metals can cause illness; however small amounts are essential to the proper functioning of many biological systems in plants and animals, including man.

Clearly, acid rain is one of the most serious international environmental problems facing our world. We recognize that coal- and oil-burning utility facilities, by dumping tons of acidic pollutants into our air every day, are major contributors to the problem, but we seem unwilling to objectively examine the alternatives. We cannot have a clean and healthy environment and continue to burn fossil fuels at the present rate. We must make a choice.

Let's set aside our instilled fears of nuclear radiation and objectively compare the relative safety of the two forms of energy generation: conventional fossil fuel and nuclear power.

In the United States, the nuclear power industry has the lowest fatality record and the lowest risk of injury in the energy generation field. Not a single member of the public has died from a commercial reactor accident! In contrast, experts estimate that coal-generated electricity results in the loss of about 10,000 lives each year, through mining and transportation accidents and respiratory illness due to higher pollution.[22] In addition to the loss of lives, we are experiencing loss of forests and lakes to the pollution from fossil fuel use.

The "accident" at Three Mile Island proved that the safety systems worked, and the disaster predicted by environmental groups and the media never materialized. The effectiveness of the safety systems is too often not given fair credit during assessments of the risks of nuclear power plants by the public. Because the public perceives coal-burning plants to be safer than nuclear power plants, public utilities continue to build new plants that require thousands of tons of coal for the generation of electricity. It takes 2.5 tons of coal compared to just 1 gram (a partial teaspoon) of uranium to produce 7,000 kilowatt-hours of electricity. Since coal-burning results in the deaths of many thousands of people each year, perhaps utilities should be encouraged to build only nuclear power plants.

In determining the risks of nuclear power, we have to evaluate the risks of the available substitutes. Clear scientific data show that alternative forms of energy production are riskier than nuclear power. The air pollution produced by a coal-burning plant is 300 times that of a nuclear power plant. Even the perception that hydroelectric plants provide clean and safe power is innaccurate, since at least 1 major dam collapses nearly every year and almost all hydroelectric generators are connected to dams. The potential death toll in California from dam failure is estimated to be 250,000 people, and statistically, the probability of a dam collapse is far greater than the probability of a nuclear power plant explosion. Besides, unless precious farmlands are flooded, nearly all the sites available for hydroelectric dams have already been developed.

America's use of electricity has grown over 45 percent since the 1973 Arab oil embargo. New plants fueled by oil will make us more dependent on oil, increase our foreign oil imports, and worsen our trade deficit, in addition to increasing atmospheric pollution and respiratory disease. New coal-powered plants will add to this pollution. Many people are unaware that nuclear energy is currently America's second largest source of electricity after coal. If we want a cleaner environment, we must reduce acid rain, slow greenhouse warming, and decrease air pollution. One step in this direction is to encourage the utility companies to build nuclear-powered energy generation plants instead of coal-burning or oil-fueled facilities.

Pesticides

Like the nuclear power industry, the pesticide industry has received a great deal of adverse publicity which is often inaccurate and exaggerates the dangers. Very recent newspaper articles, with titles like "Pesticides Pose a Hidden Threat"[23] and "Pesticides Still Sow Unknown Perils,"[24] continue to needlessly provoke public concern. We are told that pesticides are making us unhealthy, yet we are healthier than past generations, live longer, and enjoy a remarkable diversity of healthful foods which are available only through the careful use of pesticides.

"Pesticide" is the general term used to describe the class of chemicals that includes insecticides, herbicides, fungicides, and rodenticides. Pesticides are not a modern invention; they have been used since the dawn of civilization. Sulfur was used as an insecticide well before 500 B.C., and Homer wrote of its use as a fungicide.[25] The first pesticides and insecticides were derived from natural sources, and their effects have been determined by observation through the years.

By the fifteenth century, chemicals containing arsenic were being applied to crops as insecticides. Nicotine, extracted from tobacco leaves, became available as an insecticide in the seventeenth century. In the mid-1800's, two additional natural pesticides were introduced. Rotenone, discovered in extracts from the root of the derris plant, was first used by primitive tribes to paralyze fish. Derris roots were ground to a powder and then spread on the waters of lakes and rivers, and as a result, fish were paralyzed and floated to the water surface, to be collected by the hunters, who experienced no adverse effect from eating the fish. Later, after observing the high toxicity of rotenone to fish, scientists decided to evaluate the effects of the plant on insects.[26] Rotenone was shown to be an effective insecticide, and has been used on cattle and sheep to control ticks and mites, and on humans to kill head lice, scabies, and other parasites.

The other natural pesticide introduced in the mid-1800's was pyrethrum, the forefather of the still commonly used pyrethroids, which was obtained from chrysanthemum flowers. Rotenone and chemically modified pyrethrums are still in use today. The use of the ground chrysanthemum flower became a popular insecticide in many countries. By the 19th century, chrysanthemums were grown commercially in Persia, Armenia, Yugoslavia, and other countries, and in 1914, 1,500 tons of the dried flower was bought by vendors in the United

States.[27] The processing of tons of chrysanthemum flowers to obtain large volumes of the natural pesticide "pyrethrum" (so named by the Armenian flower processor Juntikoff in 1828), encouraged the growth of the American pesticide industry. Between 1920 and 1940, Japan became the world's main producer, selling about 1,800 tons of dried flowers each year. In the 1940's, Kenya, Tanzania, and Zaire increased their output of flowers and provided millions of dried crysanthemums each year.

Initially, the chrysanthemum flowers were finely ground and the powder used as the insecticide. With time, the active material in the flower was extracted and called "natural pyrethrum." It demonstrated a remarkable "knock-down" effect on flying insects and, with its low mammalian toxicity, became a favorite household insecticide. However, it was soon shown that knockdown did not necessarily result in death, as some of the flies and mosquitoes recovered. Also, natural pyrethrum is not suitable for crop use since it is rapidly destroyed by sunlight and therefore affords little insecticidal protection.

In an effort to obtain stability in sunlight, the pyrethum molecule was modified in several ways. This resulted in a class of chemical insecticides called "synthetic pyrethroids," the first of which was allethrin, introduced commercially in 1949. In the 1970's, 2 relatively stable pyrethroids, permethrin and fenvalerate, were introduced. To this day, most household insecticide sprays contain both a pyrethroid, for a quick knockdown effect, and a second insecticide such as an organophosphate, which has a slower but more lethal action. Recently, several new and improved pyrethroids, which provide superior insecticidal activity, have been introduced into the pest control market.

Let's examine some of the earliest pesticides and the development of this controversial industry. In the late 1800's, arsenic combined with copper was used to control the Colorado beetle, and in combination with lead, was used to control the gypsy moth. As the use of arsenic-containing chemicals grew rapidly, the public became more and more concerned because treated fruits and vegetables occasionally contained poisonous residues. By the late 1920's, there were sufficient numbers of poisonings and fatalities to strongly encourage a search for less toxic pesticides.

Up until this time, pesticides consisted primarily of inorganic materials – lead arsenate, copper sulfate, and other metallic compounds. With the rapid growth of the chemical industry in the 1930's, many newly synthesized organic chemicals were evaluated for pesticidal activity. In 1934, Tisdale and Williams showed the fungicidal potential of the dithiocarbamates, a group of chemicals originally developed for use as vulcanizing materials in the rubber industry. This new class of fungicide effectively controlled pathogenic fungi such as fruit rot, scabs, and potato blight. Remarkably, thiram, the first of the group to be used as a commercial fungicide, is still in use today.

In 1939 the insecticidal properties of dichlorodiphenyltrichloroethane (DDT) were discovered by the entomologist Dr. Paul Muller of the Swiss Geigy Company, during his search for an effective insecticide active against the clothes moth. DDT soon became the most widely used pesticide in the world. For his discovery, Dr. Muller received the Nobel Prize for Medicine in 1948.

Although there has been controversy in recent years over the effects of DDT on our environment, DDT has saved millions of lives. It was used during World War II, when it was applied directly to humans during military operations in the South Pacific and elsewhere to control louse-borne typhus, mosquito-carried malaria, and other diseases.[28] In 1944, to control an outbreak of typhus in Naples, Italy, more than 1,000,000 people were dusted with DDT.[29] Its introduction in India, Egypt, Thailand, and other Far East countries helped control outbreaks of malaria.[30] Therefore, the benefits of DDT were remarkable. By the mid-1950's, annual production reached 100,000 tons. More than 4 billion pounds of DDT have been used worldwide since its discovery.

There are more than 50 human diseases that are transmitted by insects.[31] Mosquitoes spread malaria, yellow fever, and encephalitis, ticks transmit spotted fever and Lyme disease; African sleeping sickness is carried by the tsetse fly; typhus and the bubonic plague (the Black Death of medieval times) are transmitted by the rat flea; and we can go on and on. While these diseases have been essentially eradicated in the United States, many are still very serious annual problems in many of the Third World countries. In those countries DDT is an important and highly effective means of controlling life-threatening diseases and is still being used.

All efforts to eradicate malaria have failed. This is partially due to our mobile society. Vietnam veterans have carried the disease back to the United States, and migrant workers apparently caused an outbreak of 27 cases in San Diego County, California, in 1986. Travellers on safaris in East Africa have in recent years developed malaria.

American physicians have the medications necessary to cure malaria, but in many Third World nations, the symptoms are either not recognized or are ignored among the poor masses. This has resulted in an alarming resurgence of malaria. For example, in 1963, only 17 cases of malaria were reported in Sri Lanka. Almost 1,000,000 cases were detected 5 years later[32]. In Africa, malaria kills more people than any other disease.[33] The Center for Disease Control in Atlanta estimates the number of cases of malaria, worldwide, to be 200 million per year.[34] As one might expect, those countries with the highest number of cases of the disease are the least able to purchase medications to treat the sick and pesticides to halt its spread. DDT is still the cheapest insecticide able to control the *Anopheles* mosquito which carries and transmits the malaria parasite. It is no wonder that DDT is still used in Africa, Asia, and other parts of the world. Certainly, a simple examination of the risks and benefits would clearly support the need to stop this deadly disease now, with DDT, without worry about the low potential of human and environmental harm which *might* be detected at some time in the future.

Despite the millions of people directly exposed to DDT, there is no evidence of direct harm to them.[35] Further, there is no recorded case of a fatal human poisoning from this insecticide.

Although extensive testing substantiates its relative safety for mammals, DDT was banned by the United States Environmental Protection Agency on January 1, 1973, not because of harmful effects to humans and other mammals, but because of environmental concerns. Some chemicals – DDT, PCB's, dioxin, and others – degrade slowly and therefore accumulate in plants and animals and in their physical surroundings. DDT is not easily metabolized by animals and consequently is stored in their fatty tissues, sometimes for many years.[36]

Numerous studies have shown that after a grain field was sprayed with DDT, the DDT was absorbed by the grain and also by the resident insect population. When birds ate the insects and the grains, they were ingesting DDT from two sources. Each time the birds ate grains and insects already containing stores of DDT, the birds added to their own body stores – this phenomenon is known as "bioaccumulation" or "biomagnification." If the birds became food for larger animals, the process of bioaccumulation continued on up through the food chain. After a rain, runoff from the treated grain field, flowing into the nearby streams and lakes, spread the DDT to life forms in the water. DDT then reached man through his meals of contaminated grain, game birds, fish, and milk and meat from cattle which had grazed in treated fields. The biomagnification process resulted in very high accumulations of DDT in man's food sources and thus in man. However, it has never been proven that accumulated DDT caused significant harm to humans,[37] although a variety of adverse effects were observed in bird and fish populations and this resulted in the banning of DDT.

DDT adversely affected calcium utilization in the hardening of bird eggs, particularly of the peregrine falcon and the bald eagle.[38] It was found that both birds had unusually high levels of DDT in their bodies, and this correlated with an inability to form sufficiently hard eggshells to sustain perpetuation of the species. DDT was also found to adversely affect game fish populations. United States government agencies and scientists agreed that the potential risks of continued use of DDT were greater than the expected benefits to our society at that particular time.

With the recognition of the broad insecticidal activity of DDT by the mid-1940's, many new molecules of similar structures were made in an effort to compete in the emerging pesticide industry. Although many hundreds were examined for insecticidal activity, very few were sufficiently active and easy enough to produce to warrant commercialization. Methoxychlor, another chlorinated insecticide introduced in the late 1940's, was shown to have an advantage over DDT because it doesn't concentrate in mammalian fat and is not persistent in the environment.[39]

At the same time, different classes of insecticides began to emerge. The organophosphate insecticides developed from research programs aimed at developing nerve gases for wartime use. While the nerve gases sarin and tabun showed excellent insecticidal activity, their high mammalian toxicity prevented their commercialization. In 1941, Schrader prepared the first real organophosphate insecticide which he called "schradan." Because of its high mammalian toxicity, Schrader synthesized additional, structurally similar molecules and in 1944, he introduced parathion, a potent insecticide still in use today.

The organochlorine-type insecticides – like DDT, aldrin, and kepone – are very stable, and therefore remain in the environment for many years and accumulate in the food chain. In contrast, the organophosphates were shown to be chemically unstable and consequently do not persist in the environment. Since the organophosphates were more environmentally acceptable, a concentrated effort was made to develop new products which do not have the significant mammalian toxicity observed with parathion, paraoxon, disulfoton, and TEPP (tetraethyl pyrophosphate). In 1950 the American Cyanamid Company introduced malathion, the first efficacious organophosphate insecticide with remarkably low mammalian toxicity. It is still widely used by commercial applicators and homeowners.

By the early 1950's, additional classes of insecticides were introduced. Since then, dozens of new insecticides have been introduced into commerce, and the chemical industry is still actively developing new insecticides which have lower mammalian toxicity and fewer environmental effects. With the right products, the farmer can attain higher crop yields while safeguarding our environment.

Rodenticides, the chemicals which kill mice, rats, and other rodents, are also classified as pesticides, and have a briefer history. Up until the introduction of warfarin in 1944, there were essentially no rodenticides having low mammalian toxicity which could selectively poison rodents. Since primitive times, rodents have spread countless diseases while invading man's home and raiding his food supplies. Even the earliest civilizations recognized rodents as carriers of highly contagious diseases such as bubonic plague and typhus.

The earliest rodenticides were nonselective, meaning they were highly toxic to both rodent and man. Strychnine was introduced in Germany in the sixteenth century as a rodenticide, and subsequently, many other poisons have been used. In 1941, Dr. Link at the Wisconsin Alumni Research Foundation was investigating why cattle often hemorrhaged and died after eating spoiled, rotting hay. He discovered the chemical responsible for these deaths to be dicoumarin, a potent anticoagulant (interferes with blood-clotting function), the ingestion of which can cause fatal internal bleeding.[40] By modifying the chemical structure of dicoumarin to obtain greater potency and stability, a series of new chemicals was synthesized. The forty-second of the series showed the greatest promise as a rodenticide and it became known as warfarin.

Warfarin acts as an anticoagulant, and the exposed rodents eventually die of internal hemorrhage. To this day, warfarin and two new "super warfarin" anticoagulants, bromdicafum and bromadiolone, are the active ingredients in almost all commercially available rodenticides. It is of interest to note that this research to develop better anticoagulant rodenticides provided the medical community with new anticoagulant drugs (medications) which are successful in the treatment and prevention of arterial blockage.

Pesticides are not just discovered and placed in the marketplace. In fact, many years of testing for both efficacy and safety is required by our federal government prior to pesticide commercialization. Information on toxicity, use rates, persistence in the environment, potential for being carried through the soil by rainwater to contaminate groundwater, effects on endangered species, maximum amount of residue expected on treated crop or shrub or lawn, and the human health risks related to the use of the material are all carefully examined by the United States Environmental Protection Agency (EPA), the federal agency which grants a registration to a pesticide manufacturer, allowing him to begin marketing his new product.[41] A registration is awarded only after all the information is reviewed by the EPA professionals and they conclude that the product is safe.

Toxic Natural Chemicals

Ideally, we would be living in a world of clean air, pure waters, and healthy foods with no form of contamination. But, we do not live in an ideal world. "Mother Nature"–through forest fires burning millions of acres per year, volcanoes blackening the skies and covering towns and villages with ash and lava, and other natural occurrences–imparts tons of chemical pollution into our air.

The most toxic chemicals known to man are made by Mother Nature, and the most potent carcinogens known are naturally occurring chemicals.[42,43] Contrary to the fables of special interest groups, nature is not benign. Nature has been making poisons for millions of years.

As animals developed, the survival of a species depended upon its ability to fend off predators and to secure food. Animals still rely heavily on their unique natural poisons. Certain snakes, spiders, and lizards readily come to mind: rattlesnakes, scorpions, the black widow spider, and the Gila monster. Several amphibians secrete poisons from glands in their own skin. These highly toxic chemicals function as a defense against predators and also help to protect the amphibian against infection by pathogenic microorganisms in the environment. These secretions effectively inhibit the growth of bacteria and fungi deadly to the animal, and when its skin is cleansed of these secretions, the animal usually dies from infection.

Not only are natural toxins important to various animal species, but they appear to be critical for the survival of plants in our environment. Scientists have shown that up to 8 percent of the dry weight of plants is toxic chemicals, formed within the plants to protect them from insect predators.[44] These are Mother Nature's own natural insecticides.

Many naturally occurring plant toxins are toxic and carcinogenic to humans.[45] The raw edible mushrooms which we enjoy in salads contain a class of natural carcinogens called "hydrazines." Hydrazines have been used as rocket propellants, herbicides, and medicinals. When a hydrazine found in mushrooms, N-methyl-N-formylhydrazine, was administered at just 0.007 percent in drinking water to rodents, it caused cancer in several organs in both mice and hamsters.[46,47] Up to 2 percent of the dry weight of raw alfalfa sprouts is canavanine, a natural chemical which when fed to monkeys, causes a serious skin disease called Lupus. Mustard contains allyl isothiocyanate, and basil contains estragole, both highly toxic naturally formed chemicals.[48] Aflatoxin B_1, the most potent carcinogen known,[49] is present in peanut butter, corn and other grains, and is also found in milk. Solanine, found in the potato plant where it protects the potato against insects, is toxic to man in amounts over 0.02 percent.[50] Solanine is not destroyed when the potato is cooked. When potatoes are exposed to sunlight, the amount of solanine in the potatoes increases at least tenfold and has frequently caused severe illness, and even human death.[51] Even the tasty tomato contains a natural insecticide, tomatine, which protects the plant against the soil borne organism that causes wilt and provides resistance to leaf spot; in people, sufficient amounts of tomatine cause alterations of cardiac contractions and disruption of red blood cells.[52]

Toxic ingredients are contained in more than 2 dozen of the herbal teas currently on the market.[53] While media attention is constantly directed to commercial pesticides, scientists have estimated that we ingest 10,000 times more natural pesticides than man-made pesticide residues.[54]

Pesticide residues and food additives are found in "organic" foods as well as in conventional foods. Some breads are claimed to be more nutritious because of the stone-ground flour, but they contain numerous sweeteners and other additives that probably cancel out any supposed nutritional bonus. In many instances, the grain is grown with the aid of pesticides and the bread will contain traces of pesticide residue. Even so, the product may be labeled "organic." The term "stone-ground" has no legal meaning. There are no universal legal definitions of "natural" or "health food"– nearly any manufacturer can use these terms to describe its products, and only a few states regulate agricultural methods for producing "organic" foods.

The current interest in "natural foods" warrants a re-emphasis on the dangers associated with natural toxins found in our foods. Remarkably, grocery shoppers pay premiums for crops grown "organically" without giving a second thought to the possible presence of natural toxins, some of which are far more potent than the majority of pesticides currently used by the farmer.

In 1960, two epidemics occurred which drew world attention. An outbreak of "Turkey-X-Disease" in England caused the death of thousands of turkeys and ducklings. Initially, officials thought the birds were poisoned, but after analyzing their peanut meal feed, the officials found the feed to be contaminated with the mold *Aspergillus flavus*. Further analysis revealed the presence of Aflatoxin B_1, a major excretory product of the mold, *Aspergillus flavus*. When aflatoxin was fed to laboratory rats, it was shown to be strongly carcinogenic, with liver tumors caused by levels as low as 15 ppb (parts per billion – a discussion soon follows which will put this measurement into perspective).

Also in 1960, a problem was discovered in the northwestern United States. Many rainbow trout, the favorite game fish of the region, had large liver tumors. An investigation by the Department of the Interior found *Aspergillus flavus* growing on the high-protein fish food that was fed to the young trout. Further analysis showed low levels of aflatoxin present on the food. The organic gardener or health food shopper might ask what finding tumors in turkeys and fish have to do with him. Plenty! Aflatoxin has been highly carcinogenic in every species in which it has been tested, including primates.

In Africa, a significant association has been found between liver tumors and aflatoxin in the diet. World Health Organization studies in Papua, Swaziland, and Kenya correlated the incidence of liver cancers with the dietary ingestion of aflatoxin. An excellent correlation was found which showed that the higher the aflatoxin levels, the greater the number of liver tumors in the local population.

More recent aflatoxin poisonings occurred in the United States. In 1977, a combination of weather and storage conditions resulted in high aflatoxin contamination of most of the corn harvested that year. Fed to farm animals, thousands of pigs and turkeys died. The milk supply in the Midwest and Southwest was contaminated because cows were unknowingly fed aflatoxin-tainted corn. Slaughterhouses found liver tumors in cattle. Consequently, the U.S. Department of Agriculture and state health departments established procedures for testing grains for aflatoxin. Additional outbreaks occurred in 1980 and 1983, resulting in greater testing and enforcement. Yet, in 1984, state regulators in Georgia forced the destruction of cornmeal found on grocery shelves in 5 pound bags, containing aflatoxin at 1000 ppb, 50 times more than the maximum permitted by the Food and Drug Administration (20ppb).

The drought of 1988 created ideal conditions for the growth of *Aspergillus flavus*, and tests of the corn harvested that year showed that almost one-third contained dangerously high levels. Many grain elevator owners destroyed their contaminated corn, but others sold it to farmers for livestock feed.

High levels of aflatoxin on corn have a direct effect on our health. The average American eats about 160 pounds of corn per year, in the form of breakfast cereals, cornmeal, tortilla chips, grits, hominy, and corn bran in health foods. If this contaminated grain was fed to cattle and dairy cows, we may be ingesting meats and milk with aflatoxin levels higher than those causing cancer in laboratory animals and fish. Since cancer takes years to develop, the impact of aflatoxin ingestion at these levels will be seen in the future. Corn isn't the only grain on which *Aspergillus* grows. It grows on peanuts, sorghum, and certain grains.

Most commercial food processors routinely test the grains they receive and reject those with excessive levels of aflatoxin. This provides us with a degree of safety not assured when purchasing grains in a health food store. If you worry about the possible health consequences of ingesting pesticide residues 1000 times *lower* than the levels which caused toxicity in animal studies, then it stands to reason that you would be very worried about eating foods contaminated with the most potent carcinogen known, in amounts *higher* than those which produce tumors in animals.

While consumer use of herbal teas is on the increase because of the perception that they are a healthful drink, Dr. Paul Ridker of the Harvard Medical School has recently described a surprising number of human poisonings resulting from the consumption of commercially available herbal teas. For example, herbal teas that contain buckhorn or senna have caused severe diarrhea and intestinal illness. Comfrey, groundsel, and gordolobo teas contain pyrrolizidine alkaloids which are toxic to the liver. These teas have caused the obstruction of critical liver veins, resulting in emergency surgery and death. Yet, because they are natural foods, they do not contain label warnings of potential side effects related to use.

Some natural products are so toxic that they have been evaluated as possible chemical warfare agents. The trichothene fungi, which are often found in moldy cereal grains, produce a chemical called T-2 toxin which, even in very small amounts, is highly irritating to the skin and mucous membranes, causing skin and oral lesions in animals and humans, resulting in severe tissue degeneration and necrosis.[55]

Every time we grill a steak or hamburger we are producing polycyclic aromatic hydrocarbons and heterocyclic amines, two types of chemicals which have been proven to cause cancer.[56] Nevertheless, we frequently take this risk voluntarily. The intake of three cancer-causing nitropyrenes found in diesel exhaust is much higher from grilled chicken than from air pollution.

Drinking water is chlorinated as part of the purification process. When the chlorine interacts with organic matter in the water, chloroform is formed in the finished water. Although the ingestion of massive amounts of chloroform by rats caused the formation of liver tumors, suggesting chloroform to be a carcinogen, there is no generally accepted substitute for the use of chlorine as a disinfectant in water. Consequently, we knowingly expose ourselves on a daily basis to chloroform.

We have read and heard that formaldehyde can cause cancer. Yet, formaldehyde is found as a natural component in almost all foods.[57] We constantly form, use, and decompose formaldehyde in our tissues as a normal part of DNA synthesis, cell division, and growth.

Nitrosamines, often associated with nitrites and used in the preservation of meats like bacon, naturally occur in certain foods, and they form in tobacco from nicotine during the curing process.[58] Nitrates and nitrites are used in the curing of processed meats primarily to provide antimicrobial protection and also to fix the characteristic pink color of processed meats. These chemicals are able to kill bacteria that cause botulism, and are the only effective anti-botulism food additive known.[59] The only alternative to nitrates and nitrites is heating the meats to very high temperatures to destroy the harmful bacteria, but, in this process, most of the heat sensitive nutrients present in the meat are also destroyed. Consequently, nitrites and nitrates are acceptable alternatives.

Many of us remember being told, in the late 1970's, that eating hotdogs, bologna, and other processed meats would place us at risk to develop cancer. Early data emerging from a large study suggested a link between the ingestion of nitrates and the development of lymphomas (cancer of the lymphatic system). Very prematurely and without the necessary validation of the study results, The Food and Drug Administration (FDA) Commissioner Donald Kennedy, in August 1978, issued a statement that nitrites had been found to pose a risk of human cancer, and that the Agency was taking steps to phase them out. Only after the announcement, and the wide publicity which it received, did the study get thoroughly reviewed by expert scientists. Their conclusion: the study was seriously flawed. There was disagreement as to the diagnosis of cancer, improper use of statistics, and a failure to analyze the test material and the animal diets for the presence of nitrosamines (a known carcinogen that is commonly found with nitrites, and can be formed by decomposition of the nitrites while under test). The FDA inspection of the laboratory while the experiment was in progress raised additional questions related to the way the test animals were handled. The National Cancer Institute independently raised questions about the diagnosis of lymphomas.

Even with this knowledge, in September 1978, the FDA still misled the public, with the statement that "We know more than enough about the Newberne study to be convinced that it is well done and strongly supports the hypothesis that nitrites are carcinogenic per se." Many of us became deeply concerned that if we fed hotdogs to our children we would be unnecessarily exposing them to a cancer-causing chemical. Fortunately, the then Secretary of Health, Education and Welfare, Joseph Califano did not accept this statement which was without documented support, and did not allow the FDA to initiate a nitrite phaseout.[60]

This recent controversy centered on the role nitrites play in the causation of cancer. While not carcinogenic themselves, nitrites can be converted to nitrosamines by interaction with protein components of meats (secondary or tertiary amines) and by interaction with the acids and digestive enzymes in our stomachs. Nitrosamines are a group of chemicals which have been shown to cause cancer in laboratory animals when administered at sufficiently high levels, daily, for long periods of time. Therefore, theoretically, if nitrosamines cause cancer and nitrites in our processed meats can be converted to nitrosamines, we are exposing ourselves to precursers of carcinogenic chemicals.

Animal tests have confirmed that daily exposure to nitrosamines in the parts per million range for long periods of time is required to cause cancer in sensitive laboratory animals.[61] Lifetime exposure to levels below one part per million did not cause cancer. While the level of exposure that is harmful to humans has not been established, the government publication, *The Cancer Bulletin*, confirms that more than 90 percent of the nitrite to which our body is exposed is produced by the body itself.[62] Reliable measurements of nitrosamines in the environment found levels in the parts per billion and parts per trillion, levels well below those acknowledged to be a risk to humans.[63]

The maximum permitted level of nitrate and nitrite in meat and fish products is 500 ppm and 200 ppm, respectively. The actual amounts of nitrates and nitrites present in processed meats has been found, in most cases, to be far below the maximum permitted level. While 500 ppm of nitrate sounds high, 600 ppm has been reported to occur naturally in cabbage and 2,000 ppm has been found in celery, spinach and beets.[64] Other vegetables, such as radishes and turnips, and several commonly eaten fish contain levels of naturally occurring nitrates almost as high as these. Since perhaps more than 70 percent of the American diet consists of vegetables and vegetable products, and cured meats account for only about 9 percent of our diet, we are ingesting many times more nitrate and nitrite from the natural constituents of our diet than from the small amounts in processed meats.

Radon

In contrast to the relatively low health risk from eating processed meats containing nitrites and nitrates, the presence of radon in our homes is a serious health hazard. Radon is a radioactive gas which forms when radium and uranium decay naturally in the ground. Both are commonly found in rocks and soil. Most soils and rocks contain from 1 to 4 ppm of uranium, and some phosphate-containing soils in Florida have levels above 100 ppm.[65] Even at these levels, when radon gas forms outdoors, it is diluted quickly and is harmlessly dispersed in the air. Outdoor radon levels are usually well below 0.4 picocuries/liter (pCi/l) of air. ("Curie" is the unit for measuring the rate of radioactive decay; a "picocurie" is one-trillionth of a curie.) However, when seeping into homes through basement floors, the gas concentrates and reaches harmful levels. Radon gas breaks down rapidly to "radon daughters" which give off highly penetrating ionizing radiation, similar to X-rays.[66] Exposure to sufficiently high levels of these radon daughters can cause cancer.

An association between radon exposure and lung cancer has been shown in radium and uranium miners who had a high incidence of lung cancer.[67] The U.S. EPA has predicted that as many as 20,000 new cases of lung cancer develop every year as a result of indoor radon pollution.[68] This is in addition to the lung cancer occurring in active smokers and passive smokers.

Radon levels in soils vary greatly, with very high levels in certain geographical areas—parts of Pennsylvania and Connecticut, Polk County in Florida, and Butte, Montana—whereas other areas contain soil and rock with relatively low levels of radon-emitting minerals.[69] In well-ventilated homes, indoor radon levels can be the same as the outdoor levels. However, the oil shortage of a decade ago and the increased cost of fossil fuels used for heating houses has encouraged homeowners to better insulate their homes, making houses as airtight as possible. Unfortunately, these energy-saving procedures keep radon from escaping and lead to a buildup of radon in many homes.[70] Radon gas diffuses into houses through basement floors, cracks in the foundation, and through open sumps; basements usually have two to three times the level found on the first or second floors of a house.

The National Research Council tells us the risk of developing cancer from exposure to ionizing radiation is proportional to the length of exposure to the radioactive material.[71] Most people spend at least 60 percent of their time indoors. Infants and the chronically ill spend up to 90 percent of their time indoors.

The average indoor radon level is 1.5 pCi/l (1.5 picocuries of radiation per liter of air). The National Council on Radiation Protection advises that continuous exposure to 4 pCi/l will result in 9 deaths due to lung cancer per 1,000 people, but the EPA Guidelines specify that radon concentrations *above* 4 pCi/l are cause for concern.[72] Adding to this inconsistency, the Connecticut Department of Environmental Protection states that long-term exposure to 4 pCi/l will cause up to 50 additional lung cancers per 1,000 population. As a comparison, a person exposed to 4 pCi/l incurs a risk equal to 200 chest X-rays per year. This is why it is important for us to have our homes tested for radon.

In energy-efficient homes in New York State, the average basement radon concentration is 6 pCi/l,[73] and in Pittsburgh, indoor levels were found as high as 24 pCi/l.[74] Northwestern states also have had high levels, the maximum recorded being 92 pCi/l.[75]

It would be wise for any house-dweller to have his basement air analyzed to determine if the radon level is sufficiently high to require remedial actions. There are remedial actions we can take to counter this natural hazard, such as sealing cracks in basement floors, increasing ventilation, and airing the house whenever possible.

Water

Groundwater

Another popular media issue is "contaminated" groundwater. We are made to feel that the major chemicals contaminating our waters are man-made. Actually, most of the chemicals found in groundwater are from natural sources and, contrary to popular opinion, the groundwater that we drink has never been "pure." The predominant contaminant is bacteria, followed by metals such as arsenic, mercury, calcium, cadmium, nickel, manganese, magnesium, and others which dissolve out of the soil and surrounding rocks directly into the water. Arsenic, mercury, and cadmium are highly toxic to humans and their presence in drinking water has caused severe health problems in families and communities relying on groundwater as their home drinking water source. Years ago, drinking water was thought to be pure because methods were not available to measure small amounts of contaminants in water. As the technology developed and sophisticated analytical methods became available, natural contaminants were also detected, since they are commonly present in higher amounts than man-made materials.

Drinking Water

When we chlorinate our drinking water to purify it, we are also causing the formation of carcinogens in the water. Is the benefit worth the risk? Most municipalities think so, and almost all public water systems in the United States are chlorinated. While the chlorine kills most of the disease-causing organisms in our water, some remain. For this reason, we occasionally learn of an outbreak of disease associated with a drinking water supply.

The purification of drinking water with chlorine is not a recently developed process. In fact, in 1908, a company that supplied drinking water derived from a sewage-contaminated reservoir to Jersey City, New Jersey, experimented with chlorinated lime solution as a disinfectant. A suit was brought against the company, requesting the court to require the more common water filtration process instead of chlorinated lime treatment. After a careful examination of the quality of the water produced by both methods, the court ruled that the water disinfected with chlorinated lime solution was as pure as filtered water.

In 1912, liquid chlorine was first used to disinfect drinking water in Niagara Falls, New York. The process was so simple and successful that by 1916, most major cities were chlorinating their water supplies. In the late 1960's, organochlorine compounds were occasionally found in drinking water.

Chlorine gas added to the water apparently combines with organic matter during the water treatment process, forming chloroform, bromodichloromethane, trichloroacetic acid, chloral hydrate, and many other chlorinated molecules. When these organochlorine compounds were first found in drinking water, regulatory agencies thought it was the result of industrial pollution. Organochlorine chemicals are widely used in the chemical, pharmaceutical, and consumer products industries, but they were found to be *not* responsible for the presence of organochlorines in the drinking water. In fact, when the EPA examined more closely the organochlorine in drinking water sources, the EPA found that it formed from the reaction of chlorine with organic materials such as decaying vegetation that were already present in the water source.

The Second International Symposium on Health Effects of Drinking Water Disinfectants and Disinfection By-Products was held in 1986 to examine whether the organochlorine compounds formed during water treatment pose a hazard to human health. If so, at what concentration and which molecules? Or, do the benefits of chlorination outweigh the risks? The conclusion reached was that while certain of the compounds were weakly mutagenic and carcinogenic at high exposure lelvels, they should not pose a significant risk to humans at the levels found in water. However, many of the products of chlorination have not been tested and one cannot say for certain that none are harmful. Certainly, on balance, the benefit appears to outweigh the risk.

The Measurement of Contaminants

The sensitivity of analytical instrumentation continues to improve. While a few years ago we marveled at being able to detect chemicals present at one part per million (ppm), and then at one part per billion (ppb), at present, many chemicals can be detected at parts per trillion (ppt). This astonishing analytical capability isn't fully appreciated by many people because they cannot relate to these numbers.

A "part per" is a fraction of something. "A part per million" (ppm) is 1 part in every 1,000,000 parts. A ppm is equivalent to 1 inch in 16 miles, 1 penny in 10,000 dollars, 1 ounce in 31 tons of potato chips, or 1 bad apple in 2,000 barrels. Clearly, a ppm is a very small amount and a material present in our foods at or below one ppm is present in very small concentrations. "A part per billion" (ppb) is 1,000 times smaller than a ppm! Therefore, a ppb is equivalent to 1 inch in 16,000 miles, 1 penny in 10,000,000 dollars, or a pinch of salt in 10 tons of potato chips. A ppb can also be looked at as 1 homerun scored in 416,667 years of major league baseball. A ppb is an incredibly small number, and yet there are those who would have us worry about exposures to chemicals at levels at or below a ppb.

The very recent development of still more sensitive analytical methods now allows us to measure certain chemicals to levels as low as a part per trillion (ppt). This is 1,000 times more sensitive than a ppb. A ppt is equivalent to 1 grain of salt in an Olympic size swimming pool full of water, or 1 grain of sand in a football field covered 18 feet high with sand. A ppt is 1 second in 320 centuries or 1 bad apple in 2 billion barrels of apples! Truly remarkable is the ability of modern analytical techniques to find that 1 part per trillion of any chemical – be is salt, chlorine, or a pesticide – in our food, air, or water. If 1 grain of sand in the football field was dyed bright red and mixed with enough sand to cover the field 18 feet deep, it would indeed be difficult to find that red grain, and the presence of 1 red grain in those billions of grains on the field surely is insignificant.

The natural carcinogen, aflatoxin, found in peanut butter and milk, is permitted to be in foods at levels as high as 20 ppb.[76] No man-made chemical product is as carcinogenic as aflatoxin. In most years, only a small portion of corn and peanut crops are contaminated. However, in 1988, widespread contamination of the corn crop was detected in the heart of the corn belt, and in states as far removed as Texas, Oklahoma, Maryland, and South Dakota. In Illinois, 34 percent of the corn samples taken from grain elevators contained levels of aflatoxin far in excess of the 20 ppb limit. Some states asked the FDA to raise the limit to 100 ppb, 5 times the maximum safe level. Since aflatoxin is a natural carcinogen, product labels do not have to inform us of the aflatoxin content. Rather than worrying about the disruption of the food supply if the aflatoxin-contaminated corn is withheld from the marketplace, federal and state agencies should address the human health risks related to the release of tainted crops. It is the amount of the toxin which determines if it will adversely affect us.

Therefore, if and when a pesticide is detected as a very small residue on a crop or in our drinking water, it is critical that we keep in perspective the amount present and consider its relative toxicity when we assess the risk. Since the quantity as well as the potency of a chemical must be considered when determining the potential risk to our health, only qualified health professionals can determine the significance of a chemical's presence based on a careful evaluation of the toxicology information available for that specific material.

Carefully conducted safety assessments are too frequently misquoted or incorrectly reported by the media. For example, the prestigious National Academy of Sciences recently examined animal test results for most of the pesticides currently used, in order to determine the cancer risk to the consumer. A day after the Academy's report was released, the Washington Post reported, "Pesticides contaminating the most common American foods may be responsible for as many as 20,000 cancer cases per year." According to Dr. Arthur Upton, a member of the committee which wrote the report, 400 cases a year would be a more accurate estimate.[77] It is indeed difficult to not be misinformed when the media coverage contains such inaccuracies.

It is of interest and great significance that there is far, far more safety assessment information available for pesticides, to which we are exposed at tiny levels, than for most of the food additives and cosmetic products to which we are exposed daily at relatively high levels, levels that are thousands of times greater than for pesticides. Perhaps one of the reasons we perceive pesticides as being harmful is because of all the publicity associated with their extensive testing. We perceive cosmetics and household products to be "safe" because we rarely hear of any adverse test results. This is not necessarily because the products are harmless. In fact, we don't know how safe their use actually is, because very little safety testing is done on them since the laws don't require the same level of testing as for pesticides. We as consumers surely are entitled to the same safeguards for *all* the foods we eat and *all* the goods we use every day.

Footnotes

1. W. F. Allman, "Staying Alive in the Twentieth Century," *Science 85*, October (1985): 31.
2. P. Slovic, "Informing and Educating the Public About Risk," *Risk Analysis* 6 (1986): 403-415.
3. J. A. Sorenson, "Perception of Radiation Hazards," *Seminars in Nuclear Medicine* (Grune and Stratton, Inc., 1986), p. 167.
4. R. A. Dobkin, *The New Reality: Update on Nuclear Plant Safety* (Atomic Industrial Forum, Bethesda, Md., 1983), p. 3.
5. *Ibid.*, p. 17.
6. The President's Commission on the Accident at Three Mile Island, *Staff Reports of the Public Health and Safety Task Force to the President's Commission on the Accident at Three Mile Island* (1979).
7. *Report of the Governor's Commission on Three Mile Island* (1980).
8. U. S. Nuclear Regulatory Commission, NUREG-0637: Report to the Nuclear Regulatory Commission from the Staff Panel on the Commission's Determination of an Extraordinary Nuclear Occurrence (ENO) *(1980)*.
9. M. Rogovan and G. T. Frampton, *NUREG/CR-1250: Three Mile Island–A Report to the Commissioners and to the Public* (Nuclear Regulatory Agency, January, 1980).
10. "Fact Sheet on the Accident at Chernobyl" (Bethesda, Maryland: Atomic Industrial Forum, 1986).
11. "Yankee Ingenuity" (Electric Council of New England, 1980).
12. K. Bridboard and J. G. French, "Carcinogenic and Mutagenic Risks Associated with Fossil Fuels," *Carcinogenesis, Vol 3: Polynuclear Aromatic Hydrocarbons*, P. W. Jones and R. I. Freudenthal, Eds. (Raven Press, 1978), pp. 451-464.
13. J. R. Louma, *Troubled Skies, Troubled Waters, The Story of Acid*

Rain (New York: Viking Press, 1984), pp. 70-71.
14. For a more comprehensive discussion of pH, see any general chemistry or general science textbook.
15. T. H. Maugh II, "Acid Rain's Effects on People Assessed," *Science* 226 (1984): 1408-1410.
16. B. G. Ferris, Jr., and J. D. Spengler, "Problems in the Estimation of Human Exposure to Components of Acid Precipitation Precursors," *Environmental Health Perspectives* 63 (1985): 5-9.
17. XPAT Reports, "Key to Reliable Exhaust Emission Data," *Environmental Science and Technology* 8 (1974): 790.
18. M. E. McDonald, "Acid Deposition and Drinking Water," *Environmental Science and Technology 19 (1985): 772-776.*
19. See: "Conference on Health Effects of Acid Precipitation," *Environmental Health Perspectives* 63 (1985): 3-174.
20. R. Kauffman, "Acid Rain, A Threat to Community Health," *Ohio State Medical Journal*, March, 1985.
21. F. A. Record and others, *Acid Rain Information Book* (Park Ridge, N. J.: Noyes Data Corp., 1981).
22. B. L Cohen and I. Lee, "A Catalog of Risks," *Health Physics* 36 (1979): 707-722.
23. Carole Sugarman, "Pesticides Pose a Hidden Threat," *Bridgeport Post*, 10 June 1987, p. 39.
24. W. J. Campbell, "Pesticides Still Sow Unkown Perils," *Hartford Courant*, 24 March 1987, p. A1.
25. R. Cremlyn, *Pesticides, Preparation and Mode of Action* (New York: John Wiley and Sons, 1978), p. 3.
26. K. A. Hassall, *The Chemistry of Pesticides* (Deerfield Beach, Florida: Verlag Chemie, 1982), p. 3.
27. *Ibid.*, p. 148.
28. K. Mellanby, *Pesticides and Pollution* (London: Collins, 1970), p. 111.
29. A. Woods, *Pest Control (London: McGraw-Hill, 1974).*
30. Ibid.
31. D. Hunter and others, *Tropical Medicine*, 5th Ed. (Philadelphia: W. B. Saunders Co., 1976).
32. E. Zamula, "Traveler's Advisory: Malaria Still Threatens Much of Globe," *FDA Consumer* 21 (1987): 8-13.
33. *Ibid.*
34. *Ibid.*
35. C. D. Klaassen and others, *Casarett and Doull's Toxicology*, 3rd Ed. (New York: MacMillan Publishing Company, 1986),

p. 543.

36. President John F. Kennedy's Scientific Advisory Committee in 1963 stated that the level of DDT in American body fat was on average, about 12 ppm, and the daily ingestion of DDT at that time was about 0.18 mg per person per day. Since then both levels have greatly diminished. By 1974, the body burden was determined to average 6.31 ppm in adepose tissue. See F. W. Kutz and others, *Pesticide Monitoring Journal* 11 (1977): 61-63.
37. W. J. Hayes, Jr., and others, "Evidence of Safety of Long-term, High, Oral Doses of DDT for Man" *Archives of Environmental Health* 22 (1971): 119-135.
38. D. Pimentel, *Ecological Effects of Pesticides on Nontarget Species* (Washington, DC: U.S. Government Printing Office, 1971).
39. K. A. Hassall, *The Chemistry of Pesticides* (Deerfield Beach, Florida: Verlag Chemie, 1982), pp 129-130.
40. W. W. Fletcher, *The Pest War* (Oxford, England: Blackwell, 1974).
41. United States Environmental Protection Agency, *The Federal Insecticide, Fungicide and Rodenticide Act of 1972, as Amended in 1978.*
42. B. N. Ames and others, "Ranking Possible Carcinogenic Hazards," *Science* 236 (1987): 271-280.
43. B. N. Ames, "Carcinogens: We Must Ignore the Trivia," *Sacremento Bee*, 21 May 1986.
44. B. Ames, "Dietary Carcinogens and Anticarcinogens," *Science* 221 (1983): 1256-1264.
45. *Ibid.*
46. B. Toth and D. Nagel, "Tumors Induced in Mice by N-methyl-N-formylhydrazine," *Journal of the National Cancer Institute* 60 (1978): 201-204.
47. B. Toth and K. Patil, "Carcinogenic Effects in the Syrian Golden Hamster of N-methyl-N-formylhydrazine," *Journal of Cancer Research and Chemical Oncology* 93 (1979): 109-121.
48. J. A. Miller and others, "The Metabolic Activation and Carcinogenicity of Alkenylbenzenes That Occur Naturally in Many Spices," *Carcinogens and Mutagens in the Environment, Vol. I*, H. F. Stich, Ed. (Boca Raton, Florida: CRC Press, 1982).
49. National Research Council, *Diet, Nutrition, and Cancer* (Washington, DC: National Academy Press: 1982), p. 237.

50. N. Sapeika, *Food Pharmacology* (Springfield, Massachusetts: Thomas Publishers, 1969), p. 67.
51. R. R. Dalvi and W. C. Bowie, "Toxicology of Solanine: An Overview," *Veterinary and Human Toxicology* 25 (1983): 13-15.
52. J. G. Roddick, "The Steroidal Glycoalkaloid, Tomatine," *Phytochemistry* 13 (1975): 9-19.
53. P. M. Ridker, "Toxic Effects of Herbal Teas," *Archives of Environmental Health* 2 (1987): 133-136.
54. B. N. Ames, "Dietary Carcinogens and Anticarcinogens," *Science* 221 (1983): 1256-1264.
55. Y. Ueno, "Trichothecenes," *Mycotoxins in Human and Animal Health*, J. Rodricks and M. A. Mehlman, Eds. (Park Forest, Illinois: Pathotox Publishers, 1977), pp. 1-4.
56. W. Lijinsky and P. Shubik, "Benzo(a)pyrene and other Polynuclear Hydrocarbons in Charcoal-broiled Meat," *Science* 145 (1964): 53-55.
57. B. N. Ames and others, "Ranking Possible Carcinogenic Hazards," *Science* 236 (1988): 271-280.
58. *Ibid.*
59. P. Magee, "Nitrosamines and Human Cancer," *Banbury Report No. 12* (Cold Spring Harbor, New York: Cold Spring Harbor Publishers, 1982).
60. R. J. Smith, "Nitrites: FDA Beats a Surprising Retreat," *Science* 209 (1980); 1100-1101.
61. G. M. Williams and J. H. Weisburger, "Chemical Carcinogens," *Casarett and Doull's Toxicology,* 3rd Ed. (New York: MacMillan Publishing Co., 1986), p. 118.
62. "Nitrates, Nitrites and Nitrosamines," *American Institute for Cancer Research Newsletter*, Volume 15 (Spring, 1987).
63. "IARC Scientific Publication No. 45" (Lyon, France: International Agency for Research on Cancer, 1983).
64. J. W. White, "Relative Significance of Dietary Sources of Nitrate and Nitrite," *Agriculture and Food Chemistry* 23 (1975): 886-891.
65. M. Eisenbud, "Sources of Ionizing Radiation Exposure," *Environment* 26 (1984): 6-33.
66. E. P. Redford, "Potential Health Effects of Indoor Radon Exposure," *Environmental Health Perspective* 62 (1985): 281-187.
67. V. E. Archer and others, "Respiratory Disease Mortality Among Uranium Miners," *Annals of the New York Academy of Sciences* 271 (1976): 280-293.
68. J. Rago, "Radon: The Risks and Remedies," *Citizens Bulletin* 14

(1987): 3-6.

69. Council on Scientific Affairs, "Radon in Homes," *Journal of the American Medical Association* 258 (1987): 668-672.
70. A. V. Nero, "Indoor Radiation Exposures from 222-Rn and Its Daughters: A Review of the Issues," *Health Physics* 45 (1983): 277-288.
71. Committee on the Biological Effects of Ionizing Radiation, National Research Council, *The Effects on Populations of Exposure to Low Levels of Ionizing Radiations* (Washington, DC: National Academy Press, 1980), pp. 37-45.
72. U.S. Environmental Protection Agency, *A Citizen's Guide to Radon: What It Is and What To Do About It* (Washington, DC: U.S. Department of Health and Human Services, 1986).
73. R. L. Fleischer and others, "Indoor Radon Levels in the Northeastern United States: Effects of Energy-efficiency in Homes," *Health Physics* 45 (1983): 407-412.
74. B. L. Cohen, "Survey of One-year Average Radon Levels in Pittsburgh Area Homes," *Health Physics* 49 (1985): 1053-1059.
75. Bonneville Power Administration, *Radon Monitoring Results from BPA's Residential Weatherization Program, Report No. 4 (Portland, Oregon: U.S. Department of Energy, 1986)*.
76. Food and Drug Administration, *Guide 7112.02, Compliance Policy Guides* (Washington, DC, 1980).
77. D. Holzman, "Cancer Risks in the Refrigerator," *Insight*, 29 June 1987, p. 52.

How Risk Is Assessed

Since small changes in a chemical's structure can result in very significant changes in its biological activity, no one can accurately predict the effect of a newly created molecule on a biological system. It is essential that newly discovered molecules, whether they be man-made or of natural derivation (extracted from plants or microorganisms), be safe when used according to label directions. Scientists have developed methods for determining a chemical's safety.

Toxicology

Toxicology is the science that specializes in determining the relative safety of materials (chemicals) with which we come in contact. Toxicologists, the scientists who conduct the safety assessments, study the adverse effects of chemicals on living systems. In recent years, with the strong public interest in recognizing and controlling hazardous chemicals, the science of toxicology has made tremendous progress in designing tests which can accurately determine the relative safety of different materials.

In order for a toxicology test to be of value, the toxicologist must be able to use the results of laboratory animal tests to predict the possible effects in humans of exposure to the same material. The toxicologist wants to derive from the animal studies answers to questions such as:

- What is the toxicity associated with the test chemical?
- At what dose (amount of exposure) does it occur and what dose is clearly without effect in the most sensitive test species?
- Is the effect reversible (goes away when exposure stops)?
- Does the effect occur in just one species or in many? (e.g., aspirin causes severe birth defects in the rat but not in other animal species.)
- What is the most sensitive exposure-related change observed in the animal that could be measured in humans exposed to the chemical?

A Case Study

A very interesting example of toxicology detective work was carried out with Ordram®, a commercial rice herbicide used worldwide to control the growth of weeds in rice paddies.[1] This account provides an excellent example of how a chemical company determined that an adverse effect observed in rodents does not occur in humans.

Ordram was evaluated in several animal species and found to be relatively nontoxic, with one exception. In a multigeneration study, in which both male and female rats received Ordram daily through two generations, Ordram caused infertility. This is a potentially serious effect. Additional tests were needed to determine if this effect, observed in rats, could occur in humans exposed to Ordram.

Since both sexes of rats had received the test chemical, scientists could not determine from this study which sex became infertile. It was not known whether Ordram caused sperm to become infertile or affected ovulation and the receptiveness of the mature egg. At the highest dose, complete infertility occurred and there were no offspring, whereas at lower doses, the effect was correspondingly less and some pregnancies occurred.

When Ordram-treated male and female rats were mated with their non-treated counterparts, the antifertility effect was shown to occur only in the male animals. At this point, the toxicologists tried to answer two questions simultaneously: (1) does the effect occur in other animal species? and (2) what is the mechanism by which Ordram causes infertility? Ordram was tested in the mouse, and while it proved to be 200 times less toxic in this species, it showed that the antifertility effect occurs in a second rodent species.

Tests needed to be conducted in other species. Ordram was then evaluated in a non-rodent species, the rabbit, which looks like a large rodent but is really a legamorph. There was no effect on fertility in the rabbit, even after the administration of massive doses for several weeks. This suggested that the antifertility effect might occur only in rodents, and therefore, humans would not be at risk. However, additional tests were needed, since this conclusion should not be based on the results from only one non-rodent species.

While Ordram's antifertility effect could have been caused by a treatment-related hormonal imbalance or damage to testicular tissue, special tests showed the effect to be on the outer membrane which holds the sperm cell together. In rodents, exposure to Ordram results in tears in the "midpiece" of the cell membrane and, in some cases, separation of the head from the tail section, making the sperm unable to fertilize an egg. Figure 5.1 shows the unusual shape of normal rodent sperm compared to sperm from rabbits, dogs, and humans, and also shows where the damage occurs.

At this point the investigators knew exactly what to look for as they evaluated Ordram in other species. Ordram was administered in massive doses for several weeks to male dogs and monkeys. When their sperm were carefully examined at

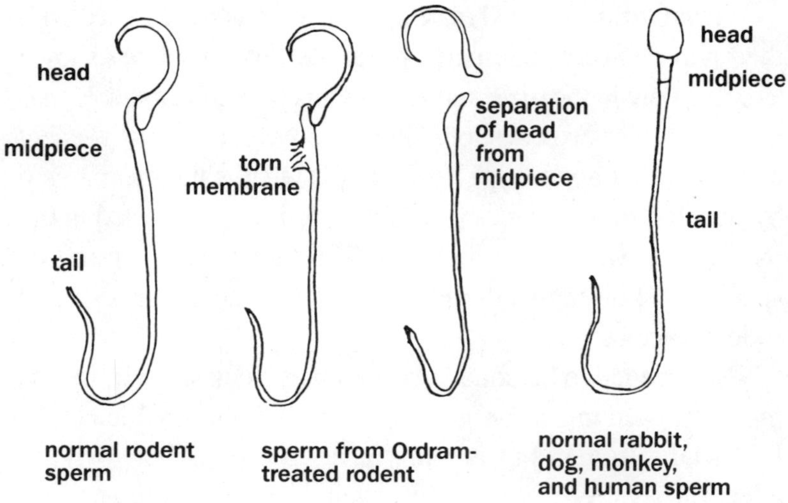

Figure 5.1: Effects of Ordram on Sperm

the conclusion of the tests using very high-powered electron microscopes, no damage was seen. Therefore, while Ordram caused infertility in male rodents by damaging sperm, it had no effect on the sperm of nonrodent species, such as rabbit, dog, and primate. As a result of these studies, the manufacturer of Ordram, the EPA, and the scientific community agreed that exposure to the pesticide did not represent a risk to man.

It is improper to test pesticides in humans, but scientists can sometimes determine the effects in workers regularly exposed at the manufacturing plant. To determine if Ordram affected the sperm of workers in the Ordram plant, the manufacturer of Ordram hired internationally recognized experts from the University of Rochester to collect and examine sperm samples from 222 male workers known to be exposed to Ordram at the manufacturing and formulating plants. The 2,187 sperm samples were collected over several months and showed no exposure-related effects. In addition, each participant completed an extensive questionaire in which his reproductive history was determined. The male workers were able to have normal, healthy children. The final report from the University of Rochester concludes that exposure to Ordram does not have a detectable effect on fertility or sperm in humans.

A question frequently asked is: why isn't Ordram, with its unique rodent antifertility properties, marketed as a rodent sterilant? It would be ineffective, because when exposure is discontinued, the antifertility effect is fully reversible as new, normal sperm mature, and the rodents are again able to bear offspring.

Toxicity

To put toxicity into perspective, we must first recognize that *all* chemicals can be toxic. Over 400 years ago, Paracelsus (a Swiss physician and alchemist) said, "All substances are poisons; there is none that is not a poison. The right dose differentiates a poison and a remedy." Since essentially all chemicals can cause injury or death under some exposure conditions, we must recognize that there is no such thing as an absolutely "safe" chemical.

Let's consider some materials which we presume to be safe. Water, essential for life, is lethal in excess; drowning results in one death out or every thirty thousand swimmers, a surprisingly high risk.[2] Excessive amounts of some vitamins, such as Vitamins A and D, can be harmful,[3] and even cause death. Oxygen, a natural component in the air and essential for human life, caused blindness in premature infants when they were placed in incubators in which the oxygen concentration was higher than in the air.[4] Because the retina of the eye is not fully developed in the newborn baby, it is exceptionally sensitive to the high oxygen concentrations once given to premature babies when it was thought to be beneficial to their growth and development. On the other hand, high levels of oxygen are routinely administered to adult patients, sports persons, and others, with beneficial results.

If many of the chemicals that we take for granted are hazardous in excessive amounts, we must determine with some degree of confidence the relative toxicity of a new material entering the marketplace as a pharmaceutical, household cleaner, cosmetic, or pesticide. Of the many types of new chemicals commercialized each year, only medications and pesticides undergo rigorous, multi-year safety assessments. Unfortunately, other new products which come into direct contact with our bodies, such as cosmetics, receive only superficial testing.

Even though we are exposed to very small amounts of pesticides, these materials receive very rigorous testing and are proven safe before they are approved by the EPA. To understand why the pesticide industry feels it is able to accurately predict potential human health effects resulting from pesticide exposure, let's briefly review the type of testing conducted and what it means to us.

The Safety Assessment

It is through multimillion dollar testing programs that the agricultural chemical industry is ascertaining the safety of currently marketed and new pesticide products. All major pesticide producers have large toxicity testing laboratories and employ many highly skilled professional toxicologists. When a new chemical is first shown in the research laboratory to be highly effective in killing weeds or insects, it is placed in a development program which simultaneously determines its commercial potential, as a herbicide or insecticide, and its toxicity. Toxicity can manifest itself in many ways, such as cancer, sterility, nerve damage, birth defects, organ degeneration, skin or eye irritation, or death. To assure that pesticides will not harm humans, domestic animals, wildlife, or the environment, a series of tests are conducted in which the new chemical is administered to a variety of laboratory animal species, simulating the routes by which humans are exposed. For example, when applying an insecticide to lawns and shrubs, one may inhale small amounts and/or get some on his skin. Also, insecticide residues on garden crops are eaten along with the produce. Therefore, the toxicity of a new pesticide is determined by administering the chemical to the animal in its diet, on its skin, and/or in the air it breathes.

Toxicity is not simply based on the administered dose of a substance, but depends partly on how much of the dose gets to the target organ. If an administered dose or level of exposure is 100 mg, only 0.1 mg may reach the target organ. For example, a chemical which may cause severe liver damage when the liver is directly exposed to milligram amounts may cause no liver damage after ingestion of 100 times that amount *if* the route of exposure is other than direct.

Therefore, to make a safety assessment of a chemical, it is necessary to determine:

(1) the rates of absorption of a chemical through the skin, lungs, and intestine
(2) the chemical's distribution to various tissues of the body
(3) whether the chemical accumulates in our tissues or is rapidly excreted
(4) the mechanism by which the chemical elicits its toxic response
(5) the minimum dose of the chemical required to cause the toxic effect.

Since new chemicals cannot be tested directly in people, special laboratory animals are used to predict toxicity in humans. Because no one species adequately mimics humans, several species are used to determine a chemical's toxicity, in order to raise the level of confidence in the safety assessment.

Acute Tests

The first tests conducted are called acute tests, in which rodents receive a single dose of the chemical, by the oral, dermal, and inhalation routes. Then usually the animals are observed for up to fourteen days to examine for lethality, skin and eye irritation, behavioral changes suggesting neurological effects, other abnormal symptoms, and tissue damage. The extremely high doses used in these acute tests are to show what could happen if one were accidentally exposed to the chemical, such as when a bag of lawn chemical breaks while being carried over the shoulder, a herbicide spray is blown into the face by a gust of wind, or an insecticide is inhaled while spraying in the home to control ants, spiders, or wasps.

Repeated-Dose Studies

To determine the possible health hazards which may result from repeated exposures to a chemical over a limited period of time, long-term exposures are designed in which four groups of animals are treated for ninety days. One group receives a very high dose, the second group gets a moderate dose, the third group gets a lower dose, and the fourth group receives no test chemical but is included as a control, representing "normal" untreated animals of that species. If the test is properly designed, significant toxicity is achieved in the top dose, mild toxicity in the mid dose animals, and no treatment-related toxicity in the low dose animals. Even the low dose animals receive considerably higher exposure to the chemical than would humans working with the material. In these ninety day subchronic tests, clinical, biochemical, and pathological examinations are conducted to determine if the test material causes an adverse effect on blood cells, blood enzymes, organ functions, and normal cellular structure and activity. If toxicity is observed, the studies may be extended to establish whether the changes persist or disappear after treatment has ceased. Microscopic examination of all major tissues by qualified pathologists assures the identification of any target organ toxicity.

Reproductive Toxicity Testing

To determine whether a chemical can affect male or female reproduction, rodents of both sexes are treated with the chemical for several weeks and are then mated. Treatment continues throughout pregnancy and nursing, and usually through two additional generations. Treatment-related effects on fertility, pregnancy, delivery, lactation, and on the growth and development of the offspring will be identified. In past years, fertility testing examined only the effects of chemicals on the female. The unfortunate occurrence of infertility in male workers manufacturing dibromochloropropane (DBCP) quickly led to an equal emphasis on male fertility.

Although a chemical may have no direct effect on reproduction, it may still express an adverse effect in the developing fetus. An abnormal change in a fetus – whether it is chemically induced or occurs as a result of hormonal imbalance, malnutrition, infectious disease, or physical trauma – is called "a teratogenic effect," with the causative agent being the "teratogen." To determine whether chemicals cause teratogenic effects in the unborn fetus, pregnant rats, mice, or rabbits are given the test chemical daily during pregnancy through the critical portion of their gestation period when the organs and limbs are developing. Just prior to delivery, the fetuses are removed and examined for skeletal, organ, or tissue abnormalities that may have developed during gestation as a result of the chemical treatment.

When evaluating for birth defects, it is important to use more than one species, since no one animal is a fully satisfactory model for humans. Aspirin, the analgesic we have come to accept as safe and efficacious, causes severe birth defects in rats but not in other species.[5] If the FDA used only rat teratology data in their safety assessment of aspirin, this valuable medication would not be on the market. In contrast, the importance of testing for teratogenic effects (birth defects caused by exposure of pregnant women to drugs and other chemicals) was not fully recognized until 1961, when the association between thalidomide ingestion by pregnant women and the birth of severely malformed infants was established. With present teratology testing methods, thalidomide would have been identified as a chemical that causes birth defects, and these deformed children would have been spared their impaired condition.

Metabolism of Chemicals

In addition to the tests just described which determine what the chemical does to the animal, metabolism tests are conducted to determine what the animal does to the chemical. When a chemical enters the body, whether it be a pesticide, medication, detergent, or environmental toxicant, the body immediately and very effectively alters the chemistry of the foreign molecule, in a process called metabolism, which takes place mainly in the liver. As described in Chapter 3, the liver, during this detoxification process, makes the chemical more water soluble so that when transported by the blood to the kidney, it will be soluble in the urine and therefore readily excreted. Otherwise, the chemical may accumulate in the body.

Metabolism studies are required to determine whether a new chemical bioaccumulates in tissues, as did DDT and as do the PCB's and dioxin, or whether it is rapidly detoxified and excreted. To be safe for humans and the environment, a chemical should not bioaccumulate in mammals, fish, plants, or soil, but should rapidly decompose to nontoxic material and eventually to the basic elements, such as carbon, oxygen, and nitrogen.

Chronic Studies

The longest tests conducted in a safety assessment are the chronic studies, which expose rodents to the test material daily for their lifetimes, which is usually 2 to 3 years. These studies, which can include as many as 800 rats or mice in a single test, examine the consequences of long-term repeated exposure on all of the body's functions and tissues. Carcinogenicity, the ability to cause cancer, is also evaluated in these lifetime studies. Here, as in the other studies described, there are at least three treatment groups (high, mid, and low doses) and a non-treated control group. For these tests to be valid, the laboratory *must* use a sufficiently high dose (called the Maximum Tolerated Dose) to produce significant toxicity, in order to learn the type of potential toxicity the material is capable of causing. Therefore, even if humans are exposed to only 0.05 ppm, the high dose for a relatively non-toxic chemical may be as high as 10,000 ppm, 200,000 times higher than actual human exposure. In this type of test, the mid and low doses might be 1,000 and 100 ppm, respectively. This lowest dose is still 2,000 times higher than the expected maximum human exposure. Also, for the test to be valid, the low dose must show no treatment-related effect, and it is then called "the no-observable-effect-level" (NOEL). Establishing the NOEL from these long-term studies is very important because the NOEL is used in risk assessments to determine safe human exposure levels for a specific chemical.

In the chronic, lifetime studies, extensive detailed records are maintained. For example, body weights and food consumption are determined weekly or bi-weekly, for each and every animal in the test. This will detect effects on growth, an early indicator of general toxicity. Clinical observations are made twice daily, seven days each week, on each animal, and these observations are recorded. Blood samples are taken at regular intervals to determine potential effects by the chemical on blood cell count, cell ratios, clotting time, blood enzymes that provide early indications of a target organ effect, and other blood parameters. The clinical laboratory making these blood measurements in a high quality toxicology facility is usually as competent and well equipped as the clinical laboratory in a first-rate hospital.

At the end of the chronic tests, all of the tissues from each animal are removed, processed, and carefully examined by a pathologist for the possible occurrence of small, subtle, treatment-related changes that can be seen only with the aid of a microscope. Therefore, if the study typically consists of 60 male and 60 female rats per dose group, and there are 3 dose groups and a control group, the study has 480 test animals. If 52 different tissue sections are examined microscopically from each animal, the pathologist will evaluate a total of 24,960 tissues before he can make his diagnosis, a task that will take about 6 months. The animal care, special diets, frequent observations, blood and tissue examinations, and daily administration of the test chemical for 2 years result in a very time-consuming, labor-intensive, and financially expensive test.

The High Cost of Toxicity Testing

At present, in 1989 dollars, the standard mammalian toxicology tests together cost over $2,000,000.00 for just one new pesticide product. These tests include: a chronic rat test which costs about $700,000, the chronic mouse test conducted simultaneously and costing about $550,000, 2 teratology tests (each about $35,000), the subchronic tests in 2 species (together about $280,000), the multigeneration reproduction test (about $280,000), metabolism ($150,000), dermal absorption ($80,000), and the single dose acute tests ($10,000). Pesticides do indeed undergo rigorous safety testing.

For pesticides, tests are also performed to determine their potential hazard to the environment, and these include tests using livestock, birds, fish, bees, beneficial insects, soil, and plants. Other tests are conducted to determine whether the chemical is able to percolate through the soil and enter groundwater, whether it degrades on the soil surface, is bound to the soil and slowly degrades, or is decomposed by soil microbes. Only if a new pesticide is shown to be environmentally safe is it awarded an EPA registration allowing it to be sold.

The combined efforts of toxicologists, chemists, ecologists, and other scientists are effectively utilized to safeguard the quality of our environment. This magnitude of testing, this multimillion dollar expense, this sincere effort to identify potential adverse health effects for new pesticides should be recognized and welcomed! It is a disservice to all of us when special interests groups downplay or misrepresent the efforts of industry and government to safeguard our health and our environment.

While the field of toxicology has made tremendous advances in the past ten years, it still cannot predict with 100 percent certainty the toxicity of every chemical. That would be as impossible as living in a completely risk-free society where people do not die from car accidents, drowning, choking on food, lightning, and other situations having prescribed risk. The use of both synthetic and natural chemicals will always be accompanied by some degree of risk, essentially the risk of living. In deciding whether to permit or reject the use of a particular chemical, we must evaluate it in terms of risk versus benefit, and we must be sensitive to the needs of society as a whole as well as to our individual preferences.

Footnotes

1. A summary of the Ordram toxicology reports can be found in: California State Water Resources Control Board Special Projects Report No. 84-4sp, Appendix 11, "Rice Herbicides: Molinate and Thiobencarb," California Toxic Substances Control Program, April, 1984.
2. R. Wilson, "Analyzing the Daily Risks of Life," *Technology Review* 81 (1979): 40-46.
3. E. M. Whelan and F. J. Stare, *The 100% Natural, Purely Organic, Megavitamin, Low-Carbohydrate Nutrition Hoax* (New York: Atheneum Press, 1984), p. 218.
4. L. Stern, "Therapy of the Respiratory Distress Syndrome," *The Pediatric Clinics of North America*, Vol. 19 (Philadelphia: W. B. Saunders Co., 1972), pp. 230-231.
5. T. Balazs, "An Overview on Delayed Toxic Effects of Pre- and Perinatal Drug Exposure," *Advances in Pharmacology and Therapeutics II* (Oxford: Pergamon Press, Ltd., 1982), pp. 163-176.

Regulations and Regulators

The chemical industry began in Europe, more than 100 years ago. At that time, the industry included leather processing, alchemy, and the manufacture of dyes and pigments. The early industry employed workers derived from the great masses. Because of inadequate diet, poor personal hygiene, the prevalence of a variety of fatal diseases, the deaths which commonly occurred at relatively early ages were thought to be of natural causes. Occupationally related disease and mortality were not considerations. However, as the chemical industry grew and the workers numbered in the thousands, correlations between occupations and illnesses were made.

The causation of cancer by occupational exposure to chemical carcinogens has been of interest in the medical community for almost 100 years. A British surgeon, Dr. Percivall Pott, was the first to identify occupationally caused cancer. In 1775, he described cancer of the scrotum as an occupational disease occurring with great frequency in chimney sweeps.[1] At that time, soot and ash were not recognized as containing carcinogenic chemicals.

In 1856, William H. Perkin synthesized pigments using aniline, and the dyes and pigments industry followed. Prior to World War I, the chemical industry in Germany supplied more than 80 percent of the world's aniline-containing chemicals, much of which was used to manufacture dyes and pigments. In 1895, Ludwig Rehn, a German physician, discovered a high incidence of bladder cancer among aniline dye workers, and by 1898 an independent report by Leichtenstern confirmed additional bladder cancers in other dye workers.[2,3] Large scale production of aniline-derived dye intermediates began in the United States during World War I, and in 1934, 27 cases of occupationally related bladder cancer were reported in the United States by Gehrmann.[4] Three years later an additional 56 cases were reported at another plant.[5]

A correlation between occupational exposure and cancer was confirmed by Hueper in 1938 when he reproduced the bladder cancer in dogs by injecting them with the aniline derivative, 2-naphthylamine, to which many of the workers had been exposed.[6] Soon after, another closely related chemical, benzidine, was shown to cause bladder cancer in laboratory animals. Although the carcinogenic properties of benzidine and 2-naphthylamine were well recognized within the chemical industry by the mid-1950's, chemical plant employees working with these materials were still not afforded adequate protection. In 1963, 11 bladder cancers were reported in workers using 2-naphthylamine in 2 small manufacturing plants in Pennsylvania[7] In 1965, 96 cases of bladder cancer were found in 366 exposed workers in New York State.[8] Additional cases were reported in the 1970's. We the public must ask why these plant workers were knowingly exposed to chemicals which were proven human carcinogens. In 1974, OSHA established "standards" for 14 carcinogens, including 2-naphthylamine and benzidine, which forced the dye and pigments industry to modify their manufacturing process, in which these chemicals are used, to assure very low workplace exposure.

Another chemical used in manufacturing, bis-chloromethyl ether (BCME), caused lung cancer in the workplace. The first cancers associated with BCME were reported in 1962.[9]

Additional BCME induced cancers were reported through 1973.[10] BCME was shown to be carcinogenic in animals subsequent to its being identified as a human carcinogen. In 1974, it was first recognized that vinyl chloride monomer, used to manufacture the commonly used plastic, polyvinylchloride (PVC), caused an unusual type of liver cancer, angiosarcoma, in humans. Subsequently, workers at several manufacturing sites using vinyl chloride developed liver cancer.[11] Studies in laboratory animals confirmed vinyl chloride as a potent carcinogen. Had the cancer been of a more common type, it may not have been correlated with chemical exposure for many additional years.

Cancer and other serious diseases occurring as a result of occupational exposure to chemicals was the driving force behind the passage of federal legislation to mandate the protection of employees through the requirement of a healthy workplace environment. New chemicals should undergo extensive animal testing before large volumes are used and workers are exposed.

We need laws which will provide reasonable assurance that our health and environment will not be impaired by the use of chemicals in the home, workplace, air, water, and foods. There exist more than a dozen federal statutes to protect our health and welfare. Let's look at some of these laws (a chronological listing is provided at the end of this chapter) and examine their importance and limitations.

The Food, Drug and Cosmetics Act

The first law to effectively control chemicals in our environment was passed in 1938, as The Food, Drug and Cosmetics Act, and after several amendments, it remains in effect. The Food, Drug and Cosmetics Act is a federal law which regulates foods, food additives, cosmetics, medical devices, and pharmaceuticals. Also, it defines, in general terms, the type of safety assessment required to bring a new food additive or pharmaceutical to market. Surprisingly, it exerts very little control over the cosmetics industry, which markets chemical products which are applied liberally and directly to the face and body as moisturizers, makeup, shampoos, and perfumes.

Prior to the passage of The Food, Drug and Cosmetics Act, the consumer had no protection from adulterated foods, harmful medications, and worthless health preparations. Even though by 1938 effective medications were available – such as aspirin for pain and fever, digitalis for the heart, barbiturates for epilepsy, and nitroglycerine for angina – ineffective and dangerous medications were also prevalent. For example, although insulin had been used since 1922 to treat diabetes, dozens of apparently useless products were also sold as cures for diabetes. The preservative sodium sulfite was commonly and illegally used in such large amounts, to disguise the smell of spoiled meats, that it caused severe kidney damage. Occasionally, raw milk from tuberculin cows was sold as pasteurized milk, accounting for as many as 5,000 deaths per year. One of the ingredients in a hair remover used by women was the rat poison, thallium acetate, which caused irreversible nerve damage. Lysol®, now sold to clean sinks and bathtubs, was first sold as a douche. Consumer reaction to such abusive practices brought about the passage of The Food, Drug and Cosmetics Act.

Two provisions of the original Food, Drug and Cosmetic Act remain as a key part of the current law. The first forbids the marketing of any food containing "any added poisonous or deleterious substance which may render it injurious to health." The second disallows the marketing of foods containing non-added toxicants which make these foods "ordinarily injurious to health." However, neither of these provisions require premarket approval from the Food and Drug Administration (FDA), the agency which administers the Act. This means a food company is able to introduce a new food product in the marketplace without FDA approval and if the agency receives complaints, the agency has the burden of proving that the food is adulterated or injurious. The Act has been modified several times by Congress, each time improving the FDA's ability to assure the safety of foods. Some of these modifications allow the FDA premarket approval of certain food substances while other changes define the types of studies necessary to obtain FDA approval.

The 1958 Food Additives Amendment

The 1958 Food Additives Amendment for substances classified as "food additives" was a very important addition to the law, as it requires that safety be demonstrated prior to marketing a new material. The law asks the manufacturer to show that the material is "reasonably certain to be safe," and the potential benefits of a new material are not considered in its safety evaluation. A major weakness in the amendment is that it does not apply to all food additives. Congress excluded certain substances which it deemed were "generally recognized as safe" (GRAS) by the qualified scientists of that time. The FDA was told to pay less attention to the GRAS substances which had been in use for many years with no observable adverse effects, but there are GRAS substances that have subsequently been proven unsafe and have been removed from the marketplace (sulfiting agents used to maintain the fresh look of salad bar ingredients, and cyclamate, the artificial sweetener, among others).

There are two types of food additives: direct and indirect. A direct or "intentional" food additive is a chemical that is added to foods to enhance flavor (monosodium glutamate), preserve quality (nitrites, BHT, BHA), enhance attractiveness (colorings), change texture (xanthan gum), or improve nutritional value (iodine to salt to prevent goiter). An indirect food additive is a chemical that enters the food unintentionally through processing or from the wrapping or container.

The Delaney Clause

The Delaney Clause was enacted in 1958 as part of the Food Additives Amendment. Written by Congressman James Delaney, this clause requires the FDA to ban any additive found to induce cancer in humans or animals. It states that *no* amount of a food additive shown to cause cancer in animals, no matter how small, will be permitted in food. It is based on three impossible standards: (1) no residue, (2) no threshold, and (3) absolute safety.

The verification of "no residue" is limited by instrumentation and methodology which, despite its sophistication and great sensitivity in measuring "ppb's" and "ppt's," cannot currently detect even smaller amounts, and therefore "no residue" is impossible to prove. "No threshold" means that for a chemical which causes cancer in animal tests only at extremely high doses, one must prove absolutely that there is no exposure concentration of the additive sufficiently low to not cause cancer. Therefore, we must assume, under this law, that even one molecule can act as a carcinogen; this concept, which disregards the body's efficient detoxification system and DNA repair capabilities, cannot be disproven. To prove "absolute safety," equivalent to zero-risk, is also an impossibility with current technology.

Another problem with Section 409 of the Delaney Clause has to do with substitute pesticides. If a non-carcinogenic substitute pesticide is available, it must be used in place of the carcinogenic product, even if the noncarcinogenic pesticide has other toxic effects. Since the Delaney Clause does not apply to noncancer effects, the FDA could disallow the presence of very small residues of a very weak carcinogen while permitting the use of a very potent neurotoxin because it is the only substitute available.

In enforcing the Delaney Clause, the FDA assumes that an additive is absent from food if it cannot be detected by current analytical methods. This is an instance where legislation has been outdated by technology—the highly sensitive analytical methods currently in use did not exist at the time the Clause was adopted. In a few cases, the FDA has allowed what it calls "negligible" amounts to be present under the *de minimis* concept, if it can be shown that the lifetime cancer risk from the trace amount present in food will not exceed one new cancer in one million people. However, for most materials, the original intent of the Delaney Clause is strictly enforced.

Unfortunately, natural carcinogens are excluded, as they are not covered by the Delaney Clause. For this reason, naturally occurring carcinogens are present in many foods we consume—aflatoxin (in peanuts and grains), polycyclic aromatic hydrocarbons (in coffee, charcoal-broiled steak, and coconut oil), pyrrolizidine alkaloids (in herbal teas), safrole (in nutmeg and cinnamon), and many others which have been identified as cancer-causing in recent research.

A 1954 amendment set limits on the amount of pesticide residue allowed in foods. For pesticides, the FDA was asked to consider both the potential adverse health effects of exposure to the pesticide and the benefits of use in deciding whether to allow a specific pesticide residue. Therefore, for pesticides, two federal agencies, the FDA and the Environmental Protection Agency provide control and protection of human health.

Regulation of Pharmaceutical Development

The introduction of a new medication requires years of preclinical animal testing for both safety and efficacy. If the preclinical studies look favorable, the pharmaceutical company asks the FDA for permission to conduct clinical trials in which the new medication is administered to people under carefully controlled conditions. The design of these clinical studies must be approved by the FDA and progress reports made at frequent intervals. Any observed side-effects are reported. These multiyear clinical tests help assure the introduction of safe new medicines. Overall, the FDA's track record is good. Although most medications pose some risk of adverse effects since they are chemicals which are designed to have biological activity, very few new medications have been introduced which have had to be withdrawn because of adverse side-effects.

Thalidomide was introduced in 1956 in West Germany as a sedative to control nausea and vomiting in pregnant women.[12] By 1960, a large increase in newborn infants of a rare limb malformation (absence of the limbs with hands evolving from the shoulders, feet from the hips) was recorded, and by 1961, an association between thalidomide use and limb malformation had been made.[13] Most of these malformed children were in Europe, with few in the United States because of the diligence of Dr. Frances Kelsey of the FDA, who, while under severe pressure from the manufacturer and from obstetricians, withheld approval of the drug pending receipt of additional safety data. Therefore, in the United States, the only use of thalidomide was limited to clinical trials in which the safety of the drug was being evaluated. In demanding additional safety data as required under federal law, the FDA saved many children from a very tragic malformation. The epidemic of limb abnormalities subsided when the drug was withdrawn from the market in late 1961, but after more than 8,000 children were deformed by thalidomide.

Non-regulated dangerous products continue to be introduced as safe and beneficial. In 1988, "herbal" diet capsules were sold through the mail as food supplements for weight loss. The label listed only herbal ingredients. However, the FDA found that the capsules contained dangerous levels of prescription drugs which could cause fluctuations in insulin levels in diabetics and seizures in epileptics, would increase blood pressure, and would cause other serious health effects. The FDA then moved to stop the sale of these dangerous capsules. Unfortunately, we will continue to be deceived by mislabeled products, so clearly, we need the protection afforded by the Food, Drug and Cosmetic Act, despite its few flaws. We need its protection and we need it to be promptly enforced.

In many areas, however, the FDA moves very slowly. This lack of action can have an adverse impact on our health. For example, 12 years ago, a public interest group asked the FDA to ban several dyes used in cosmetics and in over-the-counter medications, because cancer assessments had not been done. The Public Citizens Health Research Group founded by Ralph Nader sued the government to require safety evaluations of 84 color additives. Instead of conducting the tests, the government fought the suits and won.

In 1984, the Nader group again petitioned the FDA to test a group of dyes, identifying only 10 which they felt were potentially hazardous. Sufficient animal test data were obtained to show Red Nos. 8, 9, and 19 and Orange No. 17 to be animal carcinogens. Although federal law prohibits the use of any color additive found to cause cancer in animals or humans, the FDA took no action. Finally, in October, 1987, a federal district judge, ruling on a third lawsuit filed by the Nader group, said the FDA's reasons for keeping the colors in use were unacceptable.

The FDA maintained that human exposure was very low, and the risk to humans was correspondingly low. However, one must question whether a slight improvement in visual appearance is worth an increased risk of cancer.

One must also question the integrity of the cosmetics industry, since they continued to market products, to be applied directly to the body, which contained carcinogenic dyes and did not voluntarily substitute safer pigments in the products. It is unfortunate that the Food, Drug and Cosmetics Act does not regulate the cosmetics industry as well as it regulates pharmaceuticals and foods.

The Federal Insecticide, Fungicide and Rodenticide Act

The second oldest federal act which has significant impact on our health is the Federal Insecticide, Fungicide and Rodenticide Act (FIFRA), passed in 1947, which regulates the use of pesticides. Presently, under FIFRA, no pesticide may be marketed until it has been registered by the United States Environmental Protection Agency.[14] The law says that a pesticide may be registered if it is effective, bears proper labeling, and when properly used "will not generally cause unreasonable adverse effects on the environment." Congress asked that the benefits of use be considered along with the toxicology data.

Birth of the Environmental Protection Agency

The EPA was established in 1970 as a separate agency to administer the laws enacted by the United States Congress which protect our environment. It began its regulation of chemicals affecting our environment by quickly cancelling the registrations of several pesticides, such as DDT, aldrin, and dieldrin, based on toxicology data suggesting these chemicals may cause cancer and adversely impact our environment. This authority to quickly cancel pesticide registrations was successfully challenged by the pesticide industry and the law was revised to clarify its implementation. Congress added procedural safeguards for pesticide manufacturers and also added a panel of non-Agency expert scientists, the Scientific Advisory Panel, to provide an independent assessment of a pesticide health issue disputed by the EPA and the manufacturer. EPA established a "Special Review" procedure for the examination of serious safety issues, which includes a public hearing at which a pesticide's benefits and risks are examined. At these public meetings, all interested groups – including farmers, manufacturers, and environmentalists – are able to comment on a product's pending registrations. Under FIFRA, the EPA permits the registration and sale of a pesticide if the economic benefits from use clearly outweigh its risk to human health or the environment. If the risks are shown to be significant in this review process, the pesticide's registration is cancelled and the product removed from commerce.

The Scientific Advisory Panel

The Scientific Advisory Panel listens to all the presentations by the involved parties, including the pesticide manufacturer and the EPA, who restate their positions at this public meeting. Often the manufacturer hires independent experts as consultants to review the specific concerns raised by the EPA in order to provide an independent, objective assessment, and this is presented at the public meeting. The manufacturer, EPA, and consultants are questioned extensively on the data base used to reach their respective conclusions.

Shortly after the public meeting, the Panel provides EPA with a written opinion, explaining the basis for their conclusions, and containing recommended actions. The Panel may recommend (1) cancellation of the pesticide's registration (thereby removing it from commerce) because of demonstrated adverse health effects, (2) a reduction of the tolerance levels (the maximum residue permitted on crops), (3) the conduct of additional toxicology or environmental tests; or the Panel may agree with the manufacturer that the product is safe for its intended use and its registration will be continued as is.

While the opinion of the Scientific Advisory Panel is provided to the EPA to give guidance, the Agency is not required to abide by the recommendations of those distinguished scientists. If the EPA is not obligated to comply with the independent Panel's recommendations, there seems little value in the lengthy and costly process.

Up until 1988, the Scientific Advisory Panel was a temporary advisory board which was periodically reauthorized by Congress. In October, 1988, FIFRA Amendment S 659 was passed by Congress, authorizing several changes in the pesticide regulations. This amendment established the Panel as a permanent advisory body to the EPA.

The EPA at Work: Alar Is Reviewed

In 1984, the EPA completed a review of available toxicology data for Alar, which they interpreted as showing Alar to be a carcinogen in rodent cancer tests. Apparently, lifetime dietary ingestion of Alar and its breakdown product, UDMH, increased the tumor incidence in mice and hamsters.

Alar, also called daminozide, is a plant growth regulator, first registered by Uniroyal Chemical Company in 1968 for use on food crops. Since then, it has become an important product for farmers who grow apples, peanuts, grapes, cherries, and other fruits. For Red Delicious and McIntosh apples, it (1) increases fruit firmness, thus decreasing the number of finger bruises from picking, (2) increases storage life by 2 to 3 months, (3) delays the fruit ripening process, thereby increasing apple size and optimum red coloration, (4) controls vegetative (leaf) growth on trees, which reduces pruning costs, and (5) significantly decreases the development of "watercore," an internal rot that occurs frequently in ripe apples. It is therefore a valuable and cost-saving product to the apple grower. For the peanut farmer, Alar stimulates the upright growth of peanut plants, providing increased air circulation and light penetration, thus decreasing disease and insect problems, while increasing harvest efficiency and peanut yield.

In 1985, the Scientific Advisory Panel independently reviewed the toxicology data on which the EPA based its conclusion that Alar and its breakdown product UDMH present a cancer risk. The EPA, the manufacturer, and expert scientists gave testimony. The Panel found the older data on which EPA was basing its position to be invalid and unreliable, and therefore rejected the EPA risk estimate.

In the first long-term toxicology studies, conducted by the Food and Drug Research Laboratories in 1966, Alar was not shown to have carcinogenic activity. In 1977, Dr. Toth of the Eppley Institute published the results of a study which claimed Alar and UDMH caused cancer in laboratory rodents. Since this was the only available cancer assessment on these two chemicals, the EPA used this study as a basis for its proposed regulatory action, product cancellation.

An assessment of the Eppley studies by the Scientific Advisory Panel and independent experts found the studies so seriously flawed that they were deemed invalid. The following are some of the study deficiencies:

- The animals in the Eppley studies received Alar at doses far exceeding the MTD. The highest dose used in a toxicology study is the Maximum Tolerated Dose (MTD), a dose which causes measurable toxicity, but does not cause premature death or morbidity. The Eppley rodents received Alar at the equivalent of a human eating 50,000 pounds of Alar-treated apples per day for life.

- Only one dose was used in the Eppley studies, the dose which was above the MTD. Valid toxicology studies should consist of at least 3 treatment levels (high, mid and low doses), and a non-treated control group.
- The concurrent control animals were never fully examined to determine the normal tumor incidence in untreated rodents. Instead, the data on control animal tumor incidence reported in the Alar study was actually derived from control animals of studies conducted years earlier. Apparently, one of the control groups contained in the Eppley report was not even at the Eppley Institute. The lack of concurrent control group data for a tumor incidence comparison invalidates the entire study.
- The animals which received Alar were apparently in a state of dehydration. This had an adverse effect on the animals' health, contributed considerable stress on the biological systems, and increased the animals' vulnerability to disease and chemical insult.
- Much of the analytical data reporting Alar and UDMH concentrations were apparently inaccurate, due to the poor analytical methods used.

Recognizing these significant deficiencies, the manufacturer should have initiated its own cancer assessment tests years before the EPA Special Review. Regarding Alar, the EPA and Uniroyal are both at fault—EPA for not requiring the cancer testing in the late 1970's, and Uniroyal for not voluntarily conducting the tests so as to have definitive, valid data to support the purported safety of their product. To maintain the registration of Alar, the manufacturer was asked by the EPA to conduct new cancer assessment studies.

In 1989, the Natural Resources Defense Council (NRDC) issued a report entitled "Intolerable Risk: Pesticides in Our Children's Food," in which they claim children are at great risk of developing cancer from the use of Alar on apples. This position, unsupported by the available toxicology data, unsupported by the EPA and FDA, and unsupported by the majority of independent experts on pesticide toxicology, was given credibility through an emotional segment on the television show, "60 Minutes." The resulting public fear and hysteria has added to the challenge of determining objectively whether Alar residues on certain foods present a risk to human health, especially to the health of our children. In fact, many people have never heard of Alar and still have no idea as to why it is used.

In the preparation of their report, the NRDC used the cancer data which was rejected as being invalid by the Scientific Advisory Panel in 1985. The food consumption data obtained by the NRDC was from a small survey of only 2100 people. Their survey had a poor completion rate, with only 65 percent of the participants completing the study. In contrast, the much larger survey conducted by the United States Department of Agriculture (USDA) included over 30,000 people and had a very high completion rate of 95 percent. The USDA survey was used by the EPA in their assessment of risk.

The recent EPA risk assessment is based on preliminary results from the newest tests, just completed. Cancer assessments in rats and mice conducted at an independent testing laboratory showed that Alar did not cause cancer in either species. These results were made available to the NRDC by the manufacturer of Alar, but were not used by them in their assessment of risk. In a separate cancer test, the breakdown product UDMH caused blood vessel tumors when administered to mice at a very high dose, every day throughout their lives.

The dose was approximately 4 million times higher than the estimated human exposure. In fact, a 25 pound child would have to drink over 19,000 quarts of apple juice per day, every day for life, in order to get the same exposure to UDMH.

The NRDC report suggests that children eat more food in proportion to body weight than do adults, and because cancer tests are run in adult rodents, the results are invalid. This is not the case. Certainly, children and other sub-populations such as the aged and the infirm are more sensitive to chemical toxicity than is the general population. However, cancer tests do in fact use very young rodents which were just weaned from their mothers, and these young animals receive the test chemical every day, 7 days per week, for their entire lifetimes – through childhood, through adulthood, and through old age. Like human children, these very young rats and mice, compared to adults of these species, eat a larger amount of food in proportion to body weight and size.

How valid are the claims that residues of Alar in apple products commonly exceed the tolerance (maximum approved concentration)? In his statement to the U.S. Congress in March, 1989, FDA Commissioner Frank E. Young stated, "Over the years that we have monitored apples, apple products, and other commodities, we have yet to find a single sample in which residue from daminozide exceeded the tolerance. We rarely find levels that approach the tolerance." He went on to say that since 1981, the highest residue found was 8 ppm, less than half of the tolerance (20 ppm). It is now estimated that only 5 percent of the apple growers use Alar, compared to 25 percent in 1985.

The latest risk assessment by the EPA, based on the most recent cancer test results, showed a theoretical risk of developing cancer from lifetime exposure to Alar to be 45 additional cancers per 1 million exposed people. The EPA considers a product safe if it produces no more than 1 additional cancer per million people. While the 45 per million is not a serious risk, the EPA is proceeding to cancel the registration, and thus the sale, of Alar. Dr. John Moore, EPA Assistant Administrator, told Congress that, "While the risk estimate is greater than 1 in a million, it is not of a magnitude which dictates an imminent, overwhelming harm."

A joint statement issued by Drs. Frank Young (FDA), John Moore (EPA), and John Brode (USDA) included the following comments worth considering when reaching your own opinion on Alar. "It should be noted that risk estimates for Alar and other pesticides based on animal testing are rough and are not precise predictions of human disease. Because of conservative assumptions used by EPA, actual risks may be lower than even zero. The FDA, EPA, and USDA believe there is not an imminent hazard posed to children in the consumption of apples at this time, despite claims to the contrary."

Uniroyal Chemical, under pressure from the EPA, the apple industry, consumers, and environmental groups, held a press conference on June 2, 1989, to announce their voluntary halting of Alar sales in the United States. While Uniroyal is recalling all unsold Alar currently in the channels of trade, they still maintain that Alar is a safe pesticide. In fact, they expect the results of cancer assessments now underway to support the reintroduction of Alar for use on apples.

While national attention has been focused on the possible presence of small amounts of Alar and its degradation product, UDMH, in apples and apple juice, very little has been reported on the increased levels of the naturally occurring carcinogen, patulin, which apparently increases in the absence of Alar. Patulin, an excretory product of a fungus which grows in apples, is commonly found in apples and apple juice at levels as high as 45 ppm.[45] With the exception of aflatoxin, the level of naturally occurring carcinogens and toxins in foods is not regulated by the government. Therefore, no one is measuring the amount of patulin in apple products. Yet, with the decreased use of Alar, there will likely be an increase in patulin and other microbial toxins in apples.

Alar makes apples firmer. Without Alar, apples ripen more quickly on the tree, are less firm when ripe and blemish more easily when picked, shipped, and handled in the grocery store. Rot frequently forms in the blemish and spreads within the fruit. The rot contains fungi and other microorganisms which can excrete chemical toxins and carcinogens. While most people eat around blemished areas on fresh apples, blemished apples are sold to processors for apple juice, dried apple slices, and apple sauce. Processing and cooking does not destroy patulin, so it remains in apple products.

There is no requirement for determining either the amount or the potency of natural carcinogens in foods, thus there is little information available on the levels of patulin in apples or apple products. By discontinuing the use of Alar, we may be exposing ourselves to even greater risk from the ingestion of natural toxins whose potency is unknown.

Registration Fees

FIFRA Amendment S 659 also changed the fee structure for registering pesticides. Most people believe that we the taxpayers pay the full cost of pesticide regulation. Actually, pesticide manufacturers pay fees which cover a portion of the cost of having the EPA review their data submissions and registration applications. For example, the fee for reregistering a commercial pesticide used on crops is $150,000.

Other fees include $24,050 charged to a manufacturer whose product is being reviewed by the Scientific Advisory Panel, to help cover the cost of the Panel, and a $425 per year maintenance fee for every registered pesticide product. While this latter fee seems low, consider the following. One pesticide might be sold in several different forms: a granular solid, a dust, a water suspension, and an emulsifiable concentrate available in 3 different strengths. Those are considered to be 6 products, even though they all contain the same active ingredient, but in different formulations. The approximately 600 active ingredients sold as pesticides in the United States are included in more than 46,000 products. At $425 per product, the EPA collects in excess of 19 million dollars annually just to maintain the registration files. Many other fees are associated with registered pesticides.

Changes in Indemnification Obligations

Prior to the passage of S 659, EPA had to indemnify manufacturers, distributors, and end users of suspended or cancelled pesticides. If, because of health or environmental reasons, the EPA suspended or cancelled the registration of a pesticide, thus prohibiting its sale, the EPA became responsible for taking ownership, storing and ultimately disposing of the pesticide, and justly compensating the manufacturer, distributor, and end user for the unused goods. We, the taxpayers, were purchasing the hazardous pesticide, and then paying for its disposal.

Over the past 10 years, only 3 pesticides were removed from commerce: ethylene dibromide, Dinoseb, and 2,4,5-T/Silvex. When the EPA delayed assuming possession of the materials, several of the manufacturers entered lawsuits against the Agency. In some cases where an effective disposal method has not been identified, the EPA stored the pesticide for years, also at our expense. The 1988 Amendment S 659 limits EPA's obligation to indemnify manufacturers for unused stocks of suspended or cancelled pesticides. Under the new regulations, EPA must compensate only the end user of the cancelled pesticide, such as the farmer who might have several bags of the pesticide in his barn awaiting application to his crops. The EPA may, at its discretion, also compensate the distributor and manufacturer.

If a major manufacturer chooses to risk marketing a pesticide that borders on being unsafe, that company should be liable for all costs associated with the storage and disposal of the product, should it be suspended or cancelled. It seems unfair to shift this significant financial burden to the taxpayers.

The Clean Air Act

In 1970, the Clean Air Act was enacted. It directs the EPA to establish national air quality standards to protect human health. The Act generally regulates hazardous air pollutants and specifically regulates six major pollutants: nitrogen dioxide, sulfur dioxide, ozone, carbon monoxide, lead, and total suspended particles. In addition, the Act defines specific technology requirements that must be met, such as the installation of scrubbers on smokestacks to contain particulate and gaseous emissions.

The most well-known emission standards set by the Clean Air Act are those for automobiles, which led to the design and installation of the engine emission control systems. The Act required that the level of automobile emissions produced in 1970 be reduced by 90 percent, by no later than 1975. The introduction of catalytic converters in 1975 reduced the emissions by 83 percent. Extensions have been granted to the automotive industry for meeting the remaining 7 percent reduction requirement. A further reduction in hydrocarbons, carbon monoxide, and nitrogen oxides would be very difficult and costly.

A 1977 Amendment to the Act emphasized more stringent controls for industries burning fossil fuels. The Act places primary responsibility on state and local governments for carrying out the air pollution control programs, but gives the EPA enforcement authority, to be used when the state or local governments fail to do their parts. In fact, industry has responded positively to the Act and its amendments, and the air above many major cities is now less polluted.

The Occupational Safety and Health Act

Also in 1970, the Occupational Safety and Health Act was established to regulate exposure to toxic substances in the workplace. The Act requires that employers provide safe working conditions for employees, and it defines mandatory controls "which most adequately assure, to the extent feasible, on the basis of the best available evidence, that no employee will suffer material impairment of health or physical capacity." The law is one of the most frequently contested regulations. Many court cases have examined the language and its intent. OSHA does not regulate the use of chemicals in the workplace unless it discovers that a chemical already in use threatens the health of workers. Since OSHA cannot require toxicology testing of chemicals but must rely on existing data, and has no premarket approval authority, it is considered by many to be a weak and ineffective regulatory body.

The Consumer Product Safety Act

In 1972, the Consumer Product Safety Act was enacted which created the Consumer Product Safety Commission (CPSC) with authority to regulate products that pose an unreasonable risk of injury or illness to consumers. "Consumer products" include all items sold for use in the home, school, automobile, or elsewhere, which are not regulated by the FDA or the EPA. If a product poses an unreasonable risk of injury, the CPSC may ban the product from commerce. In recent years we have seen the CPSC regulate flame retardants in children's sleepwear, dangerous toys from which parts could be removed by babies and swallowed, the stuffing of soft dolls and animals which was microbially or chemically contaminated, and unsafe portable home heaters. The CPSC banned several substances, including vinyl chloride as a propellant, asbestos insulation, the flame retardant TRIS, and formaldehyde as a component of foam insulation.[16] However, since manufacturers are not required to notify the CPSC of plans to market a new product, the CPSC has very little real control over consumer products.

The Federal Water Pollution Control Act

The Federal Water Pollution Control Act, which originated in 1972 with subsequent amendments, and the Safe Drinking Water Act of 1974 regulate toxic water pollutants. Under the Safe Drinking Water Act, the EPA is required to regulate contaminants "which may have an adverse effect on human health" to "meet minimum national standards for the protection of public health." Control is achieved in two ways. EPA sets maximum contaminant levels for certain materials and establishes treatment procedures to control the concentration of specific toxicants in water supplies.

The Toxic Substances Control Act

One of the federal laws gaining greater importance is the Toxic Substances Control Act (TSCA) of 1976, which authorizes the EPA to require testing of chemicals which are not pesticides, food additives, drugs, or cosmetics. Prior to the enactment of TSCA, newly manufactured chemicals outside these 4 categories could be marketed without undergoing any toxicology testing.

Therefore, their ability to cause adverse effects in humans or in the environment was not determined. TSCA requires premarket notification by the manufacturer of the intention to manufacture a new chemical product and also the intended use and the estimated amounts to be made during the first three years. Although sufficient toxicology testing must be done to allow an estimate of product safety, the premanufacture notification includes very little safety data from which a reliable assessment can be made for most new chemicals. Many of the chemicals regulated by TSCA are found in the home, and there is widespread consumer exposure.

We the public should question why very little toxicology data is required for new products introduced into the marketplace in conformance with TSCA, such as flame retardants, plastics, and other specialty products, while the pesticide and pharmaceutical industries must conduct very extensive, multiyear tests to assure product safety. Why are we placed at greater risk when exposed to chemicals regulated by TSCA? The argument given by the chemical industry, which has some merit, is that, unlike pesticides and medications, for some of these specialty chemicals, the amount sold per year is so small that the profits derived cannot pay for a full complement of tests, even if every penny of profit earned were to be spent on safety evaluation. Also, for many of these specialty chemicals, there is little consumer exposure. However, the question remains: does this rationale justify the incremental risk from exposure to these inadequately tested chemicals?

A very important requirement under TSCA is the reporting to the government of any identified substantial risk related to a product. For example, if a company is allowed to market a new detergent ingredient after submitting a premanufacture notification and some simple toxicology data to the EPA, and then finds in ongoing tests that the new detergent material causes severe skin irritation, enlarged livers, or sterility in laboratory animals, the manufacturer is required to disclose this information to the EPA within thirty days or be subject to massive fines and even possible criminal charges. Unfortunately, for many chemicals regulated under TSCA, these additional safety tests are not conducted because they are not required by law. Therefore, the impact on human health may be disclosed only when and if the product causes measurable harm to those in contact with the material.

The Resource Conservation and Recovery Act

Also in 1976, the Resource Conservation and Recovery Act (RCRA) became law, regulating the transportation, treatment, storage, and disposal of hazardous wastes. The term "hazardous wastes" can be unduly broad. RCRA defines hazardous wastes as any solid, liquid, or gaseous waste material "which because of its quantity, concentration, or physical, chemical, or infectious characteristics may (1) cause or significantly contribute to an increase in mortality or an increase in serious irreversible or incapacitating reversible illness, or (2) pose a substantial present or potential hazard to human health or to the environment when improperly treated, stored, transported, disposed of, or otherwise managed." Under RCRA, the EPA must enforce the regulations applicable to waste generators, handlers, and transporters, to protect human health and the environment without consideration of economic or technical issues. In practice, both economics and technical feasibility must be carefully considered.

RCRA is one of the most important laws affecting our health and our environment. Through amendment, it was expanded in 1984 to cover hazardous materials from "small-quantity generators" (such as dry cleaners, gasoline stations, and film-processing laboratories), underground storage tanks, and other sources not previously regulated. The 1984 amendment also placed significant restrictions on the types of hazardous wastes that can be disposed of in landfills. RCRA gives the EPA broad authority over an area which we have only recently recognized as a very serious problem. By statute, the EPA must define methods for treating, storing, and disposing of hazardous wastes, and in doing so, the Agency carefully controls hazardous waste treatment, storage, and disposal facilities. RCRA requires owners to obtain permits which prescribe, in some detail, the permitted activities at that facility. Unfortunately, the combination of discovered buried wastes and the new waste being generated by various industries far exceeds the capacity of waste handlers to dispose of the materials in a safe and economically feasible manner.

The EPA is under considerable criticism for spending years studying alternative methods for disposing of hazardous wastes while wastes pile up throughout our country. New, proven technologies with extensive test data fail to receive approval. For example, the EPA has moved very slowly in awarding permits to commercial very-high-temperature incinerators which provide a relatively safe means of hazardous waste disposal.

The Comprehensive Environmental Response, Compensation and Liability Act

The transportation of hazardous materials was further regulated through passage of the Hazardous Materials Transportation Act in 1978. In 1980, the Comprehensive Environmental Response, Compensation and Liability Act (CERCLA) was passed. It allows the federal government to respond directly to releases into the environment of hazardous substances which may endanger public health. It also addresses liabilities related to chemical pollution and contamination, and provides funds for the cleanup of contaminated sites. These funds have been called "Superfund." The National Oil and Hazardous Substances Pollution Contingency Plan established the mechanism for implementing CERCLA, especially in developing action plans for assessing the risks to human health from "Superfund sites." Superfund sites are locations where large amounts of hazardous wastes have been found. Under CERCLA, the EPA has established procedures for conducting site-specific risk assessments to determine whether the hazardous waste poses a threat to human health or may result in a release into the environment.

Most of the waste sites requiring remedial action were established long before RCRA or CERCLA. Many of these sites, containing tons of buried hazardous waste, were covered over and abandoned. Being in unlined pits, many of these buried chemicals have leached through the ground and have contaminated groundwater. Under RCRA, the EPA is authorized to identify the companies which buried the wastes, and require them to pay a substantial portion of the cleanup costs. While we do not know the amount of hazardous waste generated prior to the RCRA-mandated recordkeeping, more than 500 million tons of hazardous waste was generated in 1985 alone. The Superfund program will evaluate over 320,000 sites, including 52,000 municipal landfills, 75,000 industrial landfills, 64,000 mining waste sites, and 187,000 leaking underground storage tanks.[17]

The chemical industry was primarily responsible for the abandoned buried wastes, and they are now paying for past practices. However, they were not the only one burying wastes. The leather-processing industry used large amounts of toxic metals and organics, and the pharmaceutical industry generated hazardous wastes from chemical processes used to manufacture medications. Public utilities buried ash which contains high levels of toxic metals.

A Superfund site evaluation will determine whether a contaminated site requires remedial action, such as removal of the waste for disposal elsewhere, or just properly close the site with an impermeable cap of clay. The residents of neighboring towns will be provided with alternative sources of drinking water in order to eliminate exposure via contaminated well water.

Superfund contains one additional important piece of legislation, enacted by Congress as part of the Superfund Amendments and Reauthorization Act (SARA). It is The Emergency Planning and Community Right-To-Know Act of 1986 (Title III), usually referred to simply as "Title III." This law gives communities the right to know more about chemicals used by industry within their community. Companies must adhere to 4 new requirements:

- Companies must distribute Material Safety Data Sheets to specific community organizations. These sheets must describe the chemical properties, toxicity, potential human health hazards, first aid, and disposal information for any of the hundreds of chemicals on an EPA list of toxic chemicals used at that site.
- Companies are required to inform the public of the amount (in ranges, not specific amounts) of each of the listed chemicals used, above a "threshold" amount, in the local plant. For most chemicals, a 10,000 pound per year threshold has been set, but for certain highly toxic chemicals, reporting begins at 500 pounds or less.
- Companies must report to the EPA and the community any accidental releases of listed chemicals, including the amount released into the environment.
- Each governor must appoint a state emergency response

planning committee. Companies must have emergency response plans in place by the end of 1989.

Local emergency response committees are to be established, and they are to work closely with local industry. Through Title III, better safeguards are provided for communities.

There seem to be enough laws to effectively control the generation, transport, and disposal of hazardous materials, thereby safeguarding the health of our communities. The greatest remaining risk is the large volumes of hazardous wastes buried throughout our country. While the Superfund program is addressing this urgent problem, apparently little progress is being made toward the introduction of new technologies to cope with the growing tonnage of ever-accumulating wastes. The inability of our federal agencies to take immediate and aggressive action is having a very serious impact on our environment.

Federal Legislation Regulating Chemicals

1906 *Pure Food and Drug Act*—combats food and drug adulteration and fraud (such as addition of sawdust to ground meats and franks, selling of potions with no medicinal value)

1938 *Food, Drug and Cosmetics Act (FDCA)*—regulates foods, food additives, cosmetics, medical devices, pharmaceuticals

1947 *Federal Insecticide, Fungicide and Rodenticide Act (FIFRA)*—controls the use of pesticides (was administered initially by the FDA, now by the EPA)

1954 *FDCA Amendment*—limits amount of pesticide residues allowed in foods

1958 *Food Additives Amendment (includes Delaney Clause)*—requires proof that new food additive is safe prior to addition to foods, disallows adding any amount of cancer-causing chemical to food

1970 *Environment Protection Act*—established the Environmental Protection Agency to regulate pesticides under FIFRA

1970 *Clean Air Act*—regulates hazardous air pollutants

1970 *Occupational Safety and Health Act*—regulates exposure to toxic chemicals in the workplace

1972 *Consumer Product Safety Act*—created the Consumer Product Safety Commission to regulate products posing an unreasonable risk to the consumer

1972 *Federal Water Pollution Control Act*—regulates toxic water pollutants

1974 *Safe Drinking Water Act*—regulates toxic water pollutants

1976 *Toxic Substances Control Act (TSCA)*—regulates chemicals not controlled by FIFRA or FDCA

1976 *Resources Conservation and Recovery Act (RCRA)*—regulates handling and disposal of hazardous wastes

1978 *Hazardous Materials Transportation Act*—controls transportation of hazardous materials

1980 *Comprehensive Environmental Response, Compensation and Liability Act*—defines liabilities for chemical pollution and provides funds for certain site cleanups

1984 *RCRA Amendment*—regulates "small-quantity generators," landfills, underground tanks

Footnotes

1. P. Pott, *Chirurgical Observations Relative...* (London: Hawes, Clarke and Collins, 1775).
2. L. Rehn, "Blasengeschwulste bei Fuchsin-Arbeitern," *Arch. Klin. Chir.* 50 (1895): 588-600.
3. O. Leichtenstern, "Uber Harnblasenentzundung und Harnblasengeschwulst bei Arbeitern in Farbfabriken," *Dtsch. Med. Wschr.* 24 (1898): 709-713.
4. G. H. Gehrmann, "The Carcinogenic Agent—Chemistry and Industrial Aspects," *J. Urol.* 32 (1934): 126-137.
5. E. E. Evans, "Causitive Agents and Protective Measures in the Aniline Tumor of the Bladder," *J. Urol.* 38 (1937): 212-215.
6. W. C. Hueper and others "Experimental Production of Bladder Tumors in Dogs by Administration of Beta-naphthylamine," *J. Industr. Hyg.* 20 (1938): 46-84.
7. J. Lieben, "An Epidemiological Study of Occupational Bladder Cancer," *Acta. Unio. Int.* 19 (1963): 749-750.
8. T. F. Mancuso and E. J. Coulter, "Cohort Studies of Workers Exposed to Beta-naphthylamine and Benzidine," *J. Occup. Med.* 9 (1967): 227-285.
9. L. R. Defonso and S. C. Kelton, "Lung Cancer Following Exposure to Chloromethyl Methyl Ether," *Arch. Environ. Health* 31 (1976): 125-130.
10. J. L. Creech and M. N. Johnson, "Angiosarcoma of Liver in the Manufacture of Polyvinyl Chloride," *J. Occup. Med.* 16 (1974): 150-151.
11. R. J. Waxweiler and others, "Neoplastic Risk Among Workers Exposed to Vinyl Chloride," *Ann. N.Y. Acad Sci.* 271 (1976): 40-48.
12. P. M. Fullerton and M. Kremer, "Neuropathy After Intake of Thalidomide," *British Medical Journal* 2 (1961): 855-858.

13. H. B. Taussig, "A Study of the German Outbreak of Phocomelia. The Thalidomide Syndrome," *Journal of the American Medical Association* 180 (1962): 1106.
14. When FIFRA was passed, the EPA did not exist, and the law was enforced by the FDA.
15. B. Simmonds and D. Brosten, "An Epilogue for Alar," *Agrichemical Age*, June, 1989.
16. A federal court recently overruled the CPSC action against formaldehyde, and presented a surprisingly clear and detailed opinion which pointed out the flaws in the exposure assessment used in the Commission's quantitative risk assessment.
17. D. J. Hayes and C. B. MacKerron, "Superfund II, A New Mandate," *Environmental Reporter* 17 (1987): 13.

Waste Needs Haste

Our government tells us that we in the United States generate over 150 million tons of trash each year—enough to fill 26,000 football fields, each with a layer of garbage 10 feet deep—and this amount is increasing every year.[1] Americans generate more residential trash than the citizens of any other industrialized country—more than one-half ton of trash per year for each and every man, woman, and child.

In addition to there being more trash, its content differs significantly from ten or twenty years ago, when it consisted almost exclusively of biodegradable materials. In contrast, now much of our trash is comprised of non-biodegradable plastics (containers, utensils, food wrap, automotive and computer components), synthetic fibers (clothing, carpeting, upholstery), and metal alloys which have been designed to resist the degenerative effects of the environment (stainless steel). These materials will not biodegrade when buried and will therefore remain in a landfill for generation after generation. Our enthusiastic use of disposable materials creates such an enormous amount of waste that waste management has become another very serious national issue.

Fast-Filling Landfills

The most commonly used method for waste disposal is burial

in a designated site, usually referred to as a "landfill." Over 85 percent of our trash is buried in landfills.[2] Many of the traditional landfills are already full, others have been closed because of some environmental concern, and the remaining operational landfills are rapidly reaching their limits. South Florida's last legal landfill in Broward County is almost full, yet it continues to receive more than 1,000 tons of trash daily. New York City's Fresh Kills Landfill on Staten Island receives almost 27,000 tons of trash *every day!*[3] Small wonder that we are running out of burial sites.

According to the United States Conference of Mayors, more than half of all U. S. cities will run out of landfill capacity by 1990. Some cities have already run out of space. Philadelphia pays $54 per ton to have its wastes transported to Ohio and West Virginia, and some Long Island communities pay as much as $138 per ton for out-of-state disposal.[4,5] At the rate of 27,000 tons of residential waste deposited per day in Fresh Kills Landfill on Staten Island, that 3,000 acre marshland will be filled in only 10 years.[6]

Connecticut's landfills can operate for only two more years. Already, 88 of the state's 170 cities and towns now ship their trash to other states for disposal. In 1976, New Jersey had more than 300 operating landfills. Now, there are less than 100, most small and inadequate. Twelve sites now provide space for over 90 percent of the state's garbage. Eleven New Jersey counties send their trash to facilities in other states. Both space and time are running out.

Waste disposal is an international problem. Canadian landfills are also filling up and the future disposal of waste is a pressing issue there, also. A study was conducted by the government of Canada to determine the types of waste disposed of. It found that in Toronto, disposable diapers alone accounted for 25,000 tons of the 3.4 million tons of garbage produced in that city each year.

No city or town wants a new landfill established within its boundaries. Any new landfill must be constructed to conform to government regulations, and this is expensive. Years ago, when there was little awareness of the possible health consequences of indiscriminate trash burial, any vacant quarry, mine, excavation, or wetlands became a potential landfill site. Only very recently have we become aware that chemicals can leach out of buried wastes by dissolving in rainwater, can travel through the soil, and can contaminate groundwater.

Safeguarding our groundwater (our underground supply of drinking water) is one of the most critical aspects of the waste disposal issue. Choosing the location of a new landfill site must include a hydrogeology survey to assess the distance to the closest groundwater source, an evaluation of the soil type to determine the potential for leaching to the groundwater (a clay soil retards water whereas a sandy soil allows for essentially unrestricted water movement), and other assessments.

New landfills are well-engineered collection sites. They are lined with an impermeable layer of clay or other impermeable material (such as polyvinyl chloride) to contain all deposited trash, including chemicals from the trash dissolved in rainwater. In new landfills, a system is installed at the time of landfill construction, which collects dissolved materials (household cleaners, pesticides, solvents, heavy metals from car batteries) called "leachate." This leachate either undergoes on-site neutralization or is transported to a regional treatment center. For example, let's suppose that soluble lead salts such as lead chloride are found in the leachate. In an old landfill this lead could travel with rainwater through the ground and contaminate our drinking water. When collected in the leachate, it is chemically treated to form an insoluble solid material that is placed back in the landfill. A groundwater monitoring system is installed below the newly constructed landfill to confirm the effectiveness of the impermeable barrier and that leaching is not occurring. This very simplistic diagram (Figure 7.1) shows how a new landfill is constructed.[7]

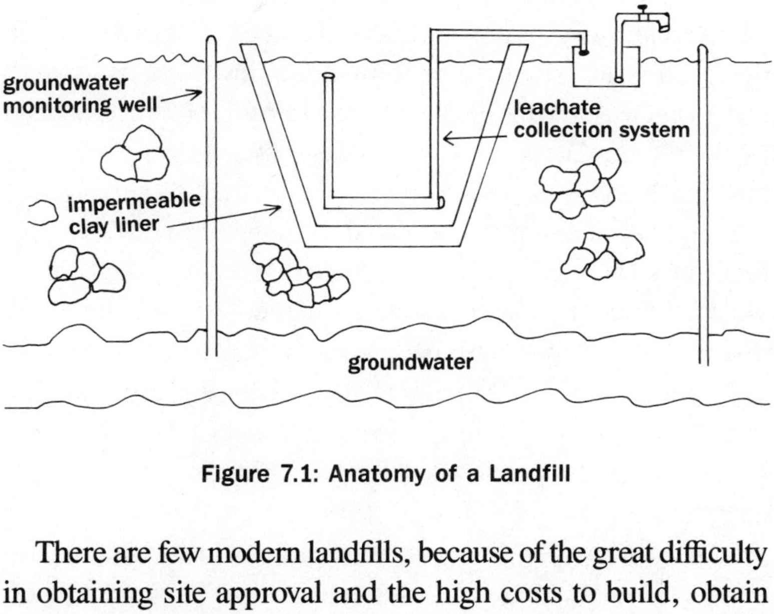

Figure 7.1: Anatomy of a Landfill

There are few modern landfills, because of the great difficulty in obtaining site approval and the high costs to build, obtain permits, operate, monitor, and close a landfill operation. Therefore, existing landfills have become a very valuable resource which must be conserved. The inescapable conclusion is that these special facilities should receive only waste that cannot be treated any other way, such as worn-out appliances and ash from commercial incinerators. Even with the clay liners and other precautions, there will always be a small risk that materials buried will someday leach to groundwater. Cracks in the clay liner could occur due to earthquakes, severe freeze-thaws, improper installation, and other factors, and thus could result in groundwater contamination. Landfills bring with them additional potential problems such as difficult-to-control underground fires and depreciation of land values in the proximity of the site.

Recycling

Environmentalists suggest that recycling is the answer to our very serious waste treatment/disposal problem. They look upon this process as a simple expansion of the Boy Scout newspaper drive or the softdrink container recycling program practiced in New York, Connecticut, and other states. In fact, recycling is a much more complex process.

Recycling can only be increased from the present 10 percent to about 20 percent of the waste stream. Areas of special effort would include: (1) raising the recycling percentage of the more than 50 billion aluminum beverage cans used each year from 50 percent to 75 or 80 percent, (2) increasing the recycling percentage of paper products (the largest single item in our waste stream) from 25 percent to 30 or 35 percent, (3) recycling approximately 5 billion glass bottles each year, and (4) searching for new ways to recycle used materials.[8] However, even recycled materials wear out and eventually must be disposed of.

A major, national recycling program would create the need for many convenient collection locations, a complex collection and transportation system, and numerous reprocessing plants. Recycling would have a negative economic impact on those businesses which supply raw materials to manufacturers, although for certain materials like aluminum, recycling makes good economic sense. It takes 20 times as much energy to obtain aluminum from bauxite ore as it does to reprocess existing aluminum materials.[9] In this case, and for glass and paper reprocessing, the savings in energy represents lower consumer costs as well as less pollution and waste from the additional fossil fuels used in the initial manufacturing process.

In a February 1988 survey by the National Solid Waste Management Association in which 1,500 adult Americans were asked about waste management, 70 percent believed that recycling "can solve much of the country's solid waste disposal problem." Yet when asked whether they would be willing to financially support a recycling program that didn't break even, only 15 percent said they would be willing to pay a direct tax or fee. Directly or indirectly through increased product costs or taxes, the consumer would pay the cost of recycling. However, even with the reuse of aluminum, steel, glass, and paper, we would still be faced with the disposal of many tons of trash that cannot be recycled.

A Better Solution

A practical, workable solution must combine burial, recycling, and incineration. Ideally, we would minimize the amount of material to be buried by recycling materials that can be efficiently reused and incinerating the remaining wastes, all at one location. At present, about 10 percent of our trash is recycled, primarily aluminum, glass, iron, and paper. As demonstrated in Europe and Japan, efficient centralized resource recovery systems could recover for recycling 20 percent or more of our residential trash. Facilities such as these are now being built.

In a commercial resource recovery incinerator system, the tons of trash received every day enter a separator which collects and separates recyclable materials, and then passes the remaining trash to a furnace. The incineration process provides sufficient energy to form steam which turns a turbine and generates electricity; 27,000 pounds of garbage (1 large garbage-truckload) produces as much energy as do 21 barrels of oil. Thus, the process converts trash to energy, and in doing so, decreases the use of fossil fuels. The sale of this electricity to the local power company helps pay for the construction and operation of the resource recovery facility (Figure 7.2).

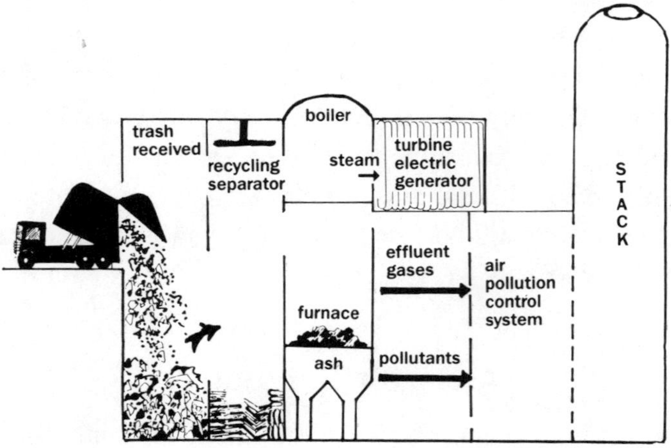

Figure 7.2: Anatomy of a Resource Recovery Facility

Not all of the new high temperature waste-to-energy incinerators being placed in operation include a recyclable materials recovery system. Some are just huge, highly efficient "mass burn" waste-to-energy units which convert all of the trash to salable steam and electricity. For many American communities, these incinerators will be the alternative to having their garbage hauled to distant landfills at incredible costs. In Florida, the Broward Landfill receives over 1,000 tons of trash per day. The 2 incinerators under construction will reduce those 1,000 tons of trash to approximately 40 tons of ash, a reduction in volume of 96 percent, with the added benefit of electricity generation in the process which, when sold, will help pay for the incinerators.[10] A waste-to-energy plant has just been completed in central Connecticut. The plant will process 12,000 tons of solid waste every week and generate enough electricity for 70,000 homes. Massachusetts has 6 operating waste-to-energy plants and 2 more are nearing completion.

The electricity generated from waste-to-energy facilities can easily be sold to support the facility. The federal Public Utility Regulatory Policy Act requires electric utility companies to purchase electricity from any business that is able to generate power for sale. The price paid by the utility companies is based on the "avoided cost" of constructing their own generating plants or buying power from another source.

The incinerator manufacturers are experiencing strong opposition in many communities. Special interest groups have challenged both the manufacturers and community leaders in court, have pressed the United States EPA to reevaluate permit applications and environmental impact studies, and have spent considerable time and money organizing uninformed residents to oppose waste-to-energy incinerators. The industry calls these organizers and their vocal converts, "nimbys" (not in my back yard). Nimbys have stopped the construction of waste-to-energy facilities in San Francisco, Seattle, Los Angeles, San Diego, and in several other cities. A one-year moratorium was just announced by the state of Massachusetts on issuing permits for additional waste-to-energy plants; thirteen permits are pending and action on those will be delayed. Unfortunately, in almost all cases, the residents of these communities did not have all the facts and options clearly laid out before them. While boldly speaking out against incineration, the nimbys offer no alternative solution, but their actions retard positive efforts toward resolving the ever-growing trash problem.

Many people seem unaware of the magnitude of the residential waste problem. We must take it upon ourselves to learn about the issues that affect our lives. We must become informed or leave it to the nimbys and other special interest groups to exploit the situation.

The Disposal of Hazardous Wastes

The disposal of hazardous wastes is another critical issue which we must understand better if we are to effectively deal with it. More and more burial sites are being unearthed, and new hazardous wastes are being formed every day. With increasing limitations on the burial of all types of wastes, very high temperature incinerative destruction appears to be the answer.

It may be easier to support a local hazardous waste incineration facility if we better understand the types of materials that would be incinerated, the strict controls on these units by federal and state regulators, and the efficiency of the disposal process. We all frequently buy, use, and transform hazardous materials into hazardous wastes. Gasoline, paints, and household cleaning solvents are classified as hazardous materials. The empty paint can, solvent container, and used paint thinner are hazardous wastes.

Used motor oil is a hazardous waste, although it is not presently listed as such.[11] In 1984, the EPA issued a Notice of Potential Risk describing a test in which mice developed skin cancer as a result of having used motor oil applied to the skin, twice a week, for about 18 months.[12] Since animal carcinogens may also cause cancer in humans, Congress directed the EPA to consider regulating used oils in order to protect the health of humans and the environment. At that time, the price of virgin fuel oil was high, and the collection and recycling of used oil as a fuel for industrial boilers and incinerators was a profitable business. However, the recent drop in virgin oil prices made oil recycling only marginally profitable, and the EPA decided that formally listing and regulating used oils as a hazardous waste, with its associated costs, would adversely impact the fragile recycling business. Therefore, even though there is evidence that used motor oil is carcinogenic in animals and contains toxic metals (arsenic, lead, cadmium) and other hazardous materials, the EPA, in August 1986, issued the following statement: "Although the EPA *proposed* to list used motor oil as a hazardous waste, we are now seriously reconsidering that proposal. Our preferred options do not involve listing recycled used oil. Used oil is *not* currently listed as a federal hazardous waste."[13] The EPA report goes on to discuss the drop in virgin oil prices and the need to maintain a recycling capability for the 700 million gallons of used oil collected each year. While this may make economic sense, it may place the thousands of people who handle used oils at an unnecessary risk. This is an instance of how we come into contact with a chemical product that should be listed as hazardous waste (because it meets the criteria for such), but is not, because of political or financial reasons.

We demand that the industrial generators of hazardous wastes conform to the very strict and comprehensive federal and state regulations governing their transport and disposal. Yet, many concerned citizens find it fully acceptable to indiscriminately dump such hazardous wastes as used motor oil and solvents into their own backyards, down the sink drain into the municipal sewage system, or into the trash for removal to the municipal landfill. Such activities pose a risk to the quality of the groundwater from which their drinking water is obtained, and could also have a significant impact on the effectiveness of the sewage treatment plant. Little bits of hazardous wastes multiplied by millions of people add up to a significant amount of illegally discarded hazardous waste. We must understand the potential problems associated with hazardous wastes and work together with government and industry to effectively protect our environment.

Wastes, which can be in solid, liquid, or gaseous form, must have one of the following characteristics to be classified as hazardous:

Ignitability—a material that may spontaneously ignite, or burn as a result of friction, presence of moisture, etc.

Corrosivity—a material which can damage or destroy other materials as in the corrosion of steel or destruction of human tissue by strong acids or caustic bases.

Reactivity—a chemically unstable material which can undergo violent change, as when pure magnesium reacts with water and bursts into flame.

Toxicity—a material which, if handled improperly, can cause a health problem.

While more than 275 million tons of hazardous waste are produced in the United States every year, only about 10 tons (less than 4 percent) are transported off-site. Under strict government control, 96 percent of hazardous waste is managed (recycled, neutralized to a non-hazardous form, or destroyed) at its source of production. The remaining 4 percent is transported by specially licensed haulers to commercial hazardous waste treatment plants for disposal. The actual method of disposal depends upon the chemical and physical characteristics of the waste. These methods include:

- Biological treatment. Certain organic chemicals can be effectively degraded by microbial digestion, forming a non-hazardous sludge that can be disposed of in a landfill or by incineration.
- Chemical treatment. Certain materials can be

neutralized by being mixed with other chemicals to decrease acidity, corrosivity, or toxicity.

- Incineration. Special incinerators capable of extremely high temperatures are able to destroy 99.999 percent of toxic waste while meeting all air emission and other environmental standards.
- Secure landfills. These are specially engineered facilities designed to contain toxic wastes and prevent leaching to the groundwater. They contain at least 3 impermeable liners and incorporate systems to collect and contain leachate and to monitor groundwater.

Hazardous waste generators are given identification numbers by the EPA; these numbers are used to track their wastes. Each and every generator is required to establish a "paper trail" which documents the type, volume, and location of the hazardous waste from the time it is generated until it is disposed of (from "cradle to grave").

We just cannot continue to bury our wastes, even hazardous wastes. The key issue is how to significantly decrease the volume of hazardous waste, and convert it to a non-toxic form, in order to safeguard our environment. High temperature incineration appears to be an answer. The design, construction, training of employee operators, and overall operation is strictly controlled at both the federal and state levels. The identification and amount of each and every hazardous waste entering the facility is recorded, the emissions constantly monitored, and the resulting ash thoroughly analyzed prior to its disposal. The safety record of these facilities is unblemished.

Since we have no place to safely store additional waste, we must support the safe destruction of hazardous wastes to protect our environment. We must first become informed if we are to provide our local and state governments with knowledgeable input concerning disposition of wastes within our community and state.

The nimbys and other activists want us to believe that hazardous waste incinerators are just built, automatically receive federal and state operating permits, and then promptly begin burning various forms of hazardous wastes. In actuality, construction entails a lengthy and complex agenda, and it can take years to obtain a permit to operate even after the facility is completed.

The ideal treatment of hazardous waste is the total destruction of the material to form just carbon dioxide, water, and certain elements. For many wastes, incineration can come very close to meeting this level of destruction. Hazardous waste incinerators are built to operate at very high temperatures, with most operating at 1,800°F. They must be designed to provide the conditions necessary for the complete oxidation (burn) of the waste, such as substantial amounts of air to supply oxygen, and some form of internal turbulence, such as a rotating cylinder, to keep the wastes mixing in the combustion chamber as they are burned.

Hazardous waste incinerators must have an EPA permit to operate. To obtain the permit, trial burn tests are conducted using the types of wastes to be destroyed at that site. To pass the test, the incinerator must destroy 99.99 percent of the waste, while controlling smokestack emissions within specific limits. The very complex incinerator control system must demonstrate full control of temperature, oxygen, and carbon monoxide concentrations, the rate of fuel use, and the accurate monitoring of stack emissions. Test burns can cost in excess of $125,000 each.

Of the 220 hazardous waste incinerators built in the United States, only 95 have received operating permits from the EPA. About 600,000 tons of hazardous wastes were incinerated in

1988. The needed capacity is about 2.5 million tons per year. While certain outspoken organizations have recently protested the permitting of hazardous waste incinerators, asking that the permits more strictly regulate certain emissions such as hydrochloric acid and metals, we must question whether we should delay operations until new technology is developed and proven. The Office of Management and Budget (OMB) and the EPA have carefully examined the potential impact of incineration on human health. The OMB concluded that hazardous waste incinerators pose less risk to an individual than does operating a riding mower.

If we delay incineration, what are our viable options? Unfortunately, only burial is currently an alternative, and the same highly vocal groups protest the burial of hazardous wastes. At present, it is more costly to incinerate hazardous waste than to bury it in a secure landfill. But, as incinerator technology improves and the facilities become more efficient, and land for new burial sites becomes even more scarce, the relative costs will reverse.

The final disposition of hazardous ash from municipal incinerators must be carefully monitored. In large residential waste incineration, the concentration of tons of waste into just pounds of ash sufficiently concentrates heavy metals in the ash to classify the ash as hazardous waste. This ash will still require regulated disposal. Some waste haulers have resorted to unethical disposal practices. In 1987, the African country of Guinea discovered 15,000 tons of waste near their capital city. Observing that trees and other vegetation around an abandoned quarry was dying, inspection of the site revealed toxic incinerator ash from Philadelphia that had been shipped and dumped by a Norwegian company. Regulations are needed which would require a dumping permit from a foreign government before waste can be loaded for shipment.

Most of the large generators of hazardous wastes handle their materials on-site, either by using them in other chemical processes or by destroying the material; therefore, the majority of the hazardous wastes to be destroyed in a regional incinerator would come from the smaller generators. These include automobile service stations, dry cleaners, photographic development laboratories, clinical (medical) laboratories, construction projects, furniture refinishers, pest control businesses, office buildings (copy machine fluids), and publishers (printing inks and dyes).

We need to identify a safe procedure for the destruction of the hazardous wastes that *we* generate daily in our homes and places of business: substances such as chlorine (used to disinfect swimming pools, etc.), naphthalene (used for moth control), 2,4-D (common broadleaf weed control herbicide sold for use on home lawns to kill dandelions, clover, etc.), vinegar, paint thinners, cleaning solvents, ether, cresol, ferrous sulfate, gasoline, and many others. Clearly, all materials classified as "hazardous" do not share the same degree of hazard – some are cumulative, but all must be disposed of properly in order to protect us and our environment.

Footnotes

1. Franklin Associates, "Characterization of Municipal Solid Waste in the United States, 1960 to 2000," Prepared for the U. S. Environmental Protection Agency, 1986.
2. *Ibid.*
3. "Tons and Tons of Trash and No Place to Put It," *U. S. News and World Report*, 14 December 1987.
4. Eliot L. Marshall, "America's Big Mess: After You Take Out the Trash, Where Will They Put It?" *Governing*, December 1987.
5. "The Rush to Burn: America's Garbage Gamble," *Newsday*, 13 December 1987.
6. Jack Lyman, "Resource Recovery: An Essential Tool for Effective Waste Management" (Washington, DC: National Solid Wastes Management Association).
7. Diagram from "Solid Waste Disposal Overview" (Washington, DC: National Solid Wastes Management Association).
8. "Recycling: Treasure In Our Trash" (Washington, DC: Waste Recyclers Council).
9. Cynthia Pollock, "Mining Urban Wastes: The Potential for Recycling" (Washington, DC: Worldwatch Institute, 1987).
10. Bill Richards, "Energy from Garbage Loses Some of Promise As Wave of the Future," *Wall Street Journal*, 16 June 1988.
11. Aspen Systems Corp., *Hazardous Waste Report*, Vol. 7 (2 September 1985).
12. U. S. Environmental Protection Agency, *Notice of Potential Risk: Used Motor Oil* (Washington, DC: February 1984).
13. U. S. Environmental Protection Agency, "Information on Used Oil Recycling for Gas Stations and Other Vehicle Service Facilities," *EPA Information Sheet* (Washington, DC: August 1986).

The "Greenhouse Effect"

"Greenhouse effect" is the term used to describe the gradual warming of our planet due to the trapping of the sun's infrared energy in the atmosphere, caused by naturally occurring and man-made chemicals. The "greenhouse effect" will adversely impact all of us, and especially the environment which our children will inherit from us. Already we are noticing its effects. There is a consistent and continuing rise in global temperature. Our summers contain more 100°+ days, we experience more water shortages, and there appear to be larger areas of drought, as in the Southeastern states which have had drought emergencies in 8 of the last 11 years. These changes are not temporary, and they will become more severe and widespread.

For millions of years, the sun has warmed the earth with its radiant heat, which is beamed to Earth as infrared energy. Much of this heat is normally reflected back through the atmosphere into space. In recent years, the increased production of carbon dioxide, nitrous oxide, and other pollutant gases has resulted in a gradual thickening of the gaseous layer in our atmosphere. While not readily seen, this layer of gases is trapping more and more of the sun's energy, allowing less to pass back into outerspace, and causing increased temperatures on earth.

If the "greenhouse" warming continues at the present rate, global temperatures will rise 3–6°C over the next century.

While this sounds like an inconsequential increase, consider that the coldest temperature during the recent Ice Age of 18,000 years ago was only about 5°C colder than it is today. It doesn't take much of a temperature change to cause global disruption and calamity.

Billions of tons of "greenhouse" gases are being added to the environment every year. Some, such as carbon dioxide, nitrous oxide, and methane, occur naturally. Additional carbon dioxide enters our atmosphere as a result of burning fossil fuels and from the burning associated with the massive deforestation in Third World countries. Each year, 50 million tons of methane and 5 million tons of nitrous oxide enter our air. In addition to contributing to the greenhouse effect, the nitrous oxide is converted to nitric acid, a component of "acid rain."

So we can expect a continued increase in global warming. By the end of this century, we will probably experience accelerated melting of glaciers, possibly raising sea level by 6 to 8 feet. This change alone could result in the flooding of populated islands, such as those in the Carribean Sea, and of coastal cities, such as New York, Miami, and San Francisco. The rising sea level would destroy wetlands and cause saltwater intrusion into rivers and shallow aquifers. But, the adverse results of atmospheric warming would not be confined to just land bound by the sea. Weather patterns would change, with the increased temperatures drying the cloud cover over the Midwest and other parts of the country, disrupting agriculture with more frequent droughts, duststorms, and higher temperatures.

We in the cities would also feel the impact. Water shortages would occur as freshwater supplies shrink through increased evaporation and decreased rainfall. In response to the higher temperatures, the increased use of electricity to cool houses, offices, and shopping centers would create more gaseous pollution from the coal-fueled and oil-burning electric-generating facilities, further escalating the greenhouse effect.

We must understand the cause of the problem and be willing to consider alternative energy sources which would slow down the generation of additional gaseous pollution. This impending crisis provides ample reason to seriously consider the use of nuclear energy to generate electricity. On October 27, 1988, Hans Blix, Director of the International Atomic Energy Agency, told the United Nations General Assembly that aside from hydropower, "nuclear power is the only source that is now available to generate electricity in the quantities, form, and reliability needed without producing any of the 'greenhouse' gases." He informed the U. N. representatives that 22 new nuclear power plants went into operation last year, bringing the world total to 417, in 26 countries. Mr. Blix urged governments to reassess their position on nuclear power in light of recent discoveries regarding climatic warming caused by the burning of fossil fuels. The reexamination of nuclear power combined with a forced reduction of deforestation and required reforestation could make a major difference.

It's hard to perceive the magnitude of our pollution. For example, it is estimated that 5 billion tons of carbon, much in the form of carbon dioxide, is released into the atmosphere, worldwide, each year. The loss of forest eliminates a critical "sink" for the carbon dioxide. A forest containing 1,000 trees would remove 6 ½ tons of carbon dioxide every year. Imagine the lost potential of the millions of trees cut every year, most of which are purchased by industrial nations such as Japan and England. Possibly, the world environmental community should impose a "use tax" on nations purchasing wood from Third World countries, which would be spent only on reforestation. How this type of tax would be imposed, and who the "keeper" of the funds would be, could be identified if the concept receives sufficient attention. We can encourage our Congress to take a stand.

How can we slow this global warming? Most nations now recognize that this is a very serious global problem. However, even if the industrialized nations take drastic steps to decrease carbon dioxide emissions, Third World countries, as they further develop, will continue to contribute more and more carbon dioxide, thereby negating the efforts of the industrialized nations. At present, the annual emission of carbon dioxide is 5 tons for every man, woman, and child in the United States, compared to about 0.3 tons in China. When China raises its per capita gross national product to only 15 percent of the U. S. level, the amount of carbon dioxide produced in China will equal current emissions from all of the coal presently burned in the United States. Significant increases in emissions will also come from India, Brazil, and the many other developing nations. Therefore, the problem must be addressed on a global level to reduce worldwide emissions.

What we hear in the media about ozone can be very confusing. We are told that ozone is harmful to our health but essential to the protection of our planet from the ultraviolet rays of the sun. Actually, both statements are correct.

The ozone formed in city smog and from burning fossil fuels is a highly reactive chemical that can damage the sensitive tissues in our lungs. In addition to causing respiratory illness, ozone adversely affects living plants. It decreases crop yields by causing serious damage. Therefore, the ozone in the lower atmosphere is harmful to our health and our environment.

In contrast, the ozone layer in the stratosphere (upper atmosphere) provides a beneficial shield, trapping the sun's ultraviolet radiation. The depletion of this protective layer of ozone by chlorofluorocarbons, nitrosoxides, carbon dioxide, and other chemicals allows more of the sun's ultraviolet radiation to reach the earth's surface, with a consequent increase in the incidence of skin cancer in humans. It also contributes to the greenhouse effect.

Much of the nitrosoxide and carbon dioxide results from the burning of fossil fuels.[1] Each year the amount progressively increases, and the damage gets worse. Again, this situation makes it imperative that we seriously consider the use of nuclear energy for electric power generation. While we will continue to produce gaseous pollutants from automobiles, the amount of total pollution can be substantially reduced through the use of alternative energy sources (nuclear, geothermic, solar) wherever feasible.

The chlorofluorocarbons, thought to deplete the ozone layer, have been widely used since the 1950's, as aerosol propellants, refrigerants, and as chemical intermediates used to make other products. They are nonflammable, noncorrosive, colorless, odorless, and chemically inert, and were therefore thought to be among the safest chemicals made by man. In 1972, the chlorofluorocarbon producers became concerned about the ultimate fate of these very stable chemicals when they were released into our environment, and a modest research program was initiated.

By 1975, attention was drawn to the chlorofluorocarbons when two researchers, Drs. Roland and Molina, introduced their hypothesis of ozone depletion by these materials.[2] Since then, much additional information has been developed which supports this hypothesis. For this reason, some uses of chlorofluorocarbons, such as a propellent gas in aerosol cans, have been banned, while other uses, such as the freon in refrigeration systems, continue until a comparable substitute chemical is identified. Additional pressure from both the government and the consumer is needed to accelerate the identification of substitute chemicals whose use would not increase our risk of skin cancer by depleting the protective ozone shield.

The combination of hydrocarbons, nitrogen oxides, and sunlight produces ozone in cities.[3] The EPA forced cities and automobile manufacturers to lower hydrocarbon emissions to decrease ozone formation. Now the government has found that trees give off large amounts of hydrocarbons, and these natural hydrocarbons are combining with nitrogen oxides to form ozone. Therefore, a mandated decrease in ozone levels may be difficult to achieve.

EPA has tried to reduce the level of industrial hydrocarbons by controlling exhaust emissions from automobiles and effluents from industrial facilities. It is estimated that in the United States, about 18 million tons of fossil-fuel-derived hydrocarbons are added to the atmosphere every year.[4] At the same time, up to 50 million tons of natural hydrocarbons are released into the atmosphere, most produced on hot summer days by trees and other vegetation, coincident with peak ozone levels. Not only do forests produce large amounts of hydrocarbons, but the types of hydrocarbons formed and emitted by trees are different than those formed from fossil fuels. The naturally formed hydrocarbons are mostly unsaturated olefins, and are much more effective in creating ozone in the lower atmosphere than the saturated hydrocarbons from motor vehicles.[5] So, while forests are essential "sinks" for the disposition of carbon dioxide, they contribute to the formation of hazardous ozone gas.

Since we cannot regulate the levels of chemicals formed by our forests, efforts should be turned to the nitrogen oxides, which combine with the hydrocarbons to form ozone. Since 1960, the number of motor vehicles in the United States has increased by 144 percent to 181 million vehicles, although the population has grown by only 33 percent to 243 million people. Many of the older automobiles still in use cause a disproportionate amount of pollution. Effective monitoring and required repairs are a must, to protect our environment. Since 1960, the number of miles driven has increased by more than 140 percent, to a staggering 1.9 trillion miles per year. This incredible number of miles driven in the U. S. helps put into perspective how motor vehicles contribute such a substantial amount of emissions pollution to our environment. A modest increase in emissions control, multiplied by the 181 million vehicles, could have a profound impact on our environment.

Several states now inspect cars annually for emissions violations. Unfortunately, many of the existing programs have little real impact. For example, the current Connecticut inspections system excludes diesel-powered automobiles and trucks. Yet diesel engines emit considerably more pollution than gasoline-fueled engines. Further, if a gasoline-fueled automobile fails the emissions inspection and the repairs necessary to pass will cost in excess of $40.00, the repairs are not required to be made and the vehicle will be excluded from further inspections. If the worst offenders are excluded, how can we hope to control the health-threatening hydrocarbons, carbon monoxide, and nitrogen oxides? The preservation of the ozone layer is important to all of us.

Footnotes

1. M. Russell, "Tropospheric Ozone and Vehicular Emissions," *Report ORNL/TM10908 (Knoxville, Tennessee: Oak Ridge National Laboratory, 1988).*
2. F. S. Rowland and M. J. Molina, "Chlorofluoromethanes in the Environment," *Reviews of Geophysics and Space Physics* 1 (1975): 1-18.
3. W. L. Chameides and others, *Science* 241 (1988): 1473.
4. P. H. Abelson, "Rural and Urban Ozone," *Science* 241 (1988): 1.
5. *Ibid.*

The Need For Balance

Everything Is Related

There are thousands of species of plants and animals, and they exist in diverse environments. Through the ecosystem, all inhabitants of the earth and their environments are related. An ecosystem is a complex system of food consumption. The ecosystem includes both living organisms and their physical environment. The earth's ecosystem includes many microcosms, or mini-ecosystems, such as ponds, deserts, and forests.

Feeding is a means of transferring energy from one life form to another. Smaller animals are consumed by larger animals, and plants are eaten by most animals. "Food chain" refers to the pathways by which nutrients and contaminants pass from species to species.

The air that we all breathe circulates around the globe, as does the water on our planet. There is continual interaction between and within microcosms to effect and maintain the balance of nature. Some processes are swift (cell division, digestion), and others move slowly (decomposition of plant and animal matter, movement of underground aquifers), but each is dependent upon other processes.

Water is necessary for the growth and development of plants and animals, and water also aids in the decomposition of plant and animal matter. In turn, the quality of the water is altered by the various plants, animals, minerals, and other matter with which it comes in contact.

Our Evolving Ecosystem

The ecosystem is always evolving and never static, and its delicate balance can be drastically upset by natural occurrences such as floods, drought, hurricanes, fires, tornadoes, volcanoes, earthquakes, and also by chemical contamination. A flood destroys natural habitats by washing away soil, nests, and other parts of the habitats. A flood kills plants and animals, and it frequently alters significantly the very composition of the area's soil. We are well aware of these and other consequences of nature's violent events.

The use of chemicals, such as insecticides, alters the balance of a microcosm. DDT has been shown to result in biomagnification, and in some areas of the world, its use has been banned. However, the current users (primarily Third World nations) consider the benefits to outweigh potential adverse environmental effects in their locales. For these nations, DDT is a valuable aid in controlling major diseases such as malaria, yellow fever, and typhus, diseases which are no longer significant health risks in the industrialized nations. Pest control remains a major problem in underdeveloped nations, as pests destroy on average about a third of all crops produced by these often undernourished populations.

Chemicals can sometimes help to restore the balance in a microcosm which has been disturbed by natural events. The careful use of appropriate pesticides can thwart an invasion of swarming locusts and prevent much of the terrible devastation they cause, and the populations of other insect pests can also be held in check with chemicals. The addition of chemicals to infertile soils can provide the nutrients necessary for healthy and high-yield crops.

Admittedly, there is danger in interfering with the balance of nature. During the Middle Ages, the Black Death killed one-fourth of Europe's population. Prior to this, there had been an intensive campaign to eliminate witchcraft by killing both witches and their "helpers," cats. Therefore, the natural mechanism for controlling the rat population was disrupted; the rats multiplied prolifically and spread the plague unchecked.

Whatever man does within his ecosystem will have an effect–sometimes great, sometimes small, sometimes harmful, sometimes beneficial. In the early 1950's, England felt it was being overrun with rabbits. A virus was identified which was highly toxic to the rabbits, and to control their rabbit population, the British spread the myxomatosis virus in 1953, and by 1955, it was estimated that about 90 percent of the rabbit population had been destroyed. Although the rabbit has made some comeback since then because of a decreased viral virulence and the development of resistance to the virus, the rabbit population remains under control.

In the early 1900's, the rabbit was introduced into Australia, as it was easy to hunt and created good sport. By the mid-1900's, the rabbit, having no natural enemy there, had overrun Australia. It ate green crops, destroyed wild vegetation, and competed with domestic animals for foods. Millions of rabbits were caught and destroyed, but they re-populated. The introduction of a lethal virus killed about 90 percent of the rabbit population. However, resistant rabbits survived and have again re-populated Australia, and they continue to destroy wild vegetation by eating the roots, causing serious erosion. In Australia, the purposeful addition of a new species has had very serious consequences.

Technological advances often have unanticipated effects. In

1829 the Welland Canal was completed, allowing ships to bypass Niagara Falls and sail from the Atlantic Ocean through the Great Lakes and into the Upper Midwest of the United States. The Welland Canal also allowed marine animals to migrate from the Atlantic into the Great Lakes, and the sea lamprey took advantage of this opportunity. This peculiar fish has a strange mouth with which it attaches itself to the body of any large, healthy fish, which it eventually kills by sucking out the host's vital juices. It took many years for the sea lamprey to effectively migrate into the Great Lakes. In the early 1900's, trout were bountiful in the Lakes – in 1935, fishermen landed 6.6 million pounds of trout from the U. S. side of Lake Huron and from Lake Michigan, and in 1940, it was 7.1 million, but by 1949 the lampreys were flourishing and only 300,000 pounds of trout were caught. The lampreys had almost wiped out the trout population in the Great Lakes.

Canadian and U. S. scientists began working together to develop means for controlling the lampreys. When electric gates in spawning sites did little good, they turned to chemicals for a solution. They needed to find one that would selectively kill the lamprey without injuring other water creatures. In 1955, a material referred to as TFM was shown to kill lampreys but not other fish. Carefully conducted field trials showed that TFM, used to kill lamprey, would not be a health hazard to humans nor have an adverse impact on the environment. By 1961 the lamprey had been eliminated and the trout population recovered, aided by the introduction of hatchery-produced fish.

It is well recognized that while the industrialized countries produce crops in great excess, large segments of the world population go hungry. The starvation and associated high mortality rate in Third World countries emphasizes the urgent need for innovative technologies that will support the agricultural practices which have some potential for success in these regions. In many of the Third World countries, agricultural production competes unsuccessfully with one or more environmental or man-made factors, such as drought, infertile soils, insect and weed infestations, insufficient training, and the use of seeds which are inappropriate for the region.

New products are presently being developed to selectively control weeds and insects identified as problems in industrialized countries, with little consideration given to their applicability to the special agricultural problems identified within some developing nations. Yet, those emerging countries cannot afford the costs associated with the independent development of pesticides specific for their region. Global efforts are needed to assure that new, safer, non-persistent, region-specific pesticides are also developed to meet this special need. These new products could dramatically enhance crop yields in those regions now experiencing starvation because of inadequate crop production. With improved crop yields, the agricultural communities of the developing nations could move forward to self-sustaining agricultural production.

Biotechnology could have a substantial, positive impact on sustainable agricultural ecology. This highly specialized science is still in its infancy, but its potential is overwhelming. New technologies are emerging which could have a major, positive impact on world agriculture. Several crops are being genetically modified to develop new strains which are resistant to specific insects. Most crops are attacked by crop-specific insects. For example, the corn borer and corn rootworm account for almost all of the insect damage occurring in corn. Some laboratories are developing modified corn plants which will be able to produce specific insect toxins, which will make the plant highly resistant to the crop-specific insect. This insect toxin manufactured by the corn plant would give sufficient crop protection from insect infestation to provide commercially acceptable yields without the use of chemical insecticides. Very recently, insect resistance has been engineered into cotton plants, providing resistance to cotton bollworm and other pests. A new protein produced by the modified cotton plant interferes with the insects' feeding ability.

The incorporation of specific genetic material into crops to impart insect resistance should also decrease and possibly eliminate the need for crop-specific chemical insecticides. This is especially important in underdeveloped nations where both costs and potential health hazards associated with the use of insecticides remain an issue. Insect-resistant crops may also be successful in areas previously considered inappropriate for agriculture because of high insect infestation. The development of special crop varieties for Third World countries would require costly multi-year research programs, and there is controversy concerning who should provide the financial support for such a project.

The area of biotechnology also includes the development of genetically modified microorganisms such as virus and bacteria which can function as target-specific biological pesticides. For example, rather than spraying large regions or bodies of water with persistent synthetic insecticides like DDT to eliminate or control disease vectors such as mosquitoes, a microorganism might be developed which would infect only mosquitoes, would have a short natural life span, and would infect no other host. This would assure applicator safety, provide minimal risk to exposed populations, and have significant global value. Obviously, great control must be exercised during the development of genetically modified organisms to assure minimal potential for accidental exposure during development or premature release before its safety is confirmed. However, developed properly, these organisms would offer the advantages of host specificity, lack of toxicity to other forms of life, and would not leave chemical residues on crops or pollute drinking water. They therefore would have a positive environmental impact. Hopefully, scientific ingenuity and emerging technologies will provide other approaches to global sustainable agricultural ecology.

Our environment is a depletable resource. According to a 1987 Worldwatch Institute Report, Brazil, Indonesia, and Zaire contain 48 percent of the world's tropical forests. While these forests use enough carbon dioxide during photosynthesis to significantly offset the greenhouse effect, these countries are allowing private deforestation at an incredible rate with a loss of almost a million acres of forest per week.

In 1987, Brazil cleared an area half the size of Texas! To quickly clear some of the land, the local people are burning square miles of the Amazon forest. This burning contributes 200 billion tons of carbon dioxide per year to the environment, 10 percent of the world production of the harmful gas.

Some Conclusions

Because of the magnitude of the burning, this must be considered a world problem, since it is affecting our world environment, and could have a devastating effect on the world we leave to our children. Unfortunately, the burning continues with no world organization pressing for a responsible reduction. Every day counts! We need to make our concerns known.

Not only are we gradually destroying the air we breathe and the stratospheric ozone which affords us protection from cancer-causing ultraviolet rays, we are also affecting the groundwater essential to our future. In the United States, more than 50 million people obtain their drinking water from underground aquifers. People throughout the world depend on rivers and lakes for their drinking water, yet we continue to contaminate it.

We deplete our soil of nutrients through overuse and poor cultivation practices. Added to man's effects is the substantial depletion of our environmental resources through natural occurrences—floods, forest fires, storms, droughts, and earthquakes.

While we look upon our natural resources as renewable, they cannot be regenerated at the pace at which they are currently being consumed. Therefore, we must develop a more responsible relationship with our environment. We must question the value of practices that needlessly deplete a resource or endanger the survival of a species. For example, perhaps we should voluntarily stop using styrofoam cups, plastic bags, and other items which do not decompose in our environment, even though this will cause some inconvenience.

The recognition that a human is a mixture of chemicals, and that our entire universe is composed of chemicals, makes the term "chemical" less ominous. Since we are unwilling to return to the lifestyles of 100 years ago, we have to realistically and objectively consider the relative risks pertaining to our twentieth-century lifestyles. A willingness to make modest changes in our use of chemical products can have a significant positive effect on the environment.

Driving from New York City to the Newark Airport, one can't not notice the huge municipal landfill which appears as an isolated mountain, its height increasing hourly from the continuous dumping of the garbage trucks that seem tiny in comparison. How high is high enough? With more than half of our nation's landfills scheduled to be full by the early 1990's, now is the time to consider alternative waste disposal methods. It takes years to locate a site, obtain federal and state permits, design, construct, and begin operation of a safe and efficient resource recovery incinerator. With no other practical alternative available, it seems reasonable to begin planning now, rather than when our landfills are full.

As we write these words on this hot, summer afternoon, our desklights dim again. The TV news-announcer describes power outages in several New York City locations and in parts of Connecticut. Newspapers headline the worsening energy shortages which threaten the region's economic growth. The public power supply company asks us to use less electricity—to reset the thermostats on climate-control systems, turn off excess lights, and avoid using appliances unnecessarily—because of power shortages and overloads. Sometimes, this same power company says that their extra capacity for generating electricity means additional generators are not needed. Yet, every year, the cost of electricity increases significantly, even as service is decreased just at the times when it is most needed.

Many dollars are spent trying to modify existing energy generation facilities to comply with federal and state effluent-discharge regulations. Taller smokestacks, the addition of "scrubbers" to collect particulates, and changes in the design and temperature of furnaces still result in the accumulation of tons of ash and particulate matter which must be hauled away and disposed of. Where to and at what cost? This appears to be the time for a re-examination of nuclear-powered electric generation which is clean, generates little waste, is safe, and will not exceed current emission standards.

Even with all of the media attention given to pesticides, it is reassuring to know that these chemicals undergo extensive testing for both human and environmental safety. It is further reassuring to learn that after carefully comparing the animal and human toxicology data available for forty-four chemicals, the Director of EPA's Office of Health and Environmental Assessment just announced that the animal data overstated the risk to humans by a factor of ten.[1] Further, the EPA found that the animal and human data were "highly correlated" and provided added confidence to the current safety assessment programs.

Most of us appreciate the availability of a diverse selection of fresh fruits and vegetables almost year-round, made possible by the use of safe pesticides. Without the availability of these chemicals, the diminishing number of farmers would not be able to supply the nation's food. Only 1.6 percent of the U.S. population is engaged in farming compared to 71 percent of the Chinese, 37 percent of Mexicans, 34 percent of Russians, and 13 percent of Japanese.[2] Yet, "naturalists" suggest banning pesticides and growing gardens in every backyard. How many New Yorkers or Chicagoans have backyards? How many have the time or skills for cultivation?

In 1910 the average American farmer produced enough food for about 7 people, but today the average farmer feeds 80 people. And the food is less expensive now. In the United States we spend about 13 percent of our gross income for food, compared to 60 percent in China, 41 percent in Mexico, 31 percent in Russia, and 22 percent in Japan. In addition to helping provide an abundance of healthy food, pesticides allow us to spend less for our food. This gives us more of our spendable income for housing, clothing, transportation, medical care, and education, with some left over for luxuries.

Our ability to detect very small amounts of chemicals has advanced much faster than our ability to determine the meaning and significance of the findings. Some experts suggest that the potential risk from low level exposure is greatly exaggerated. More basic research is needed to better understand the body's chemistry and metabolic processes and its interaction with foreign chemicals. We must cease to be fearful of the possible risk from some minute amount of a substance which is carcinogenic to rodents. Test animals are exposed to extremely large amounts for a lifetime, whereas our exposure is in very small amounts, intermittently. Therefore, the high-dose animal studies may not accurately predict effects of low-dose human exposure but may, in fact, significantly exaggerate the actual risk.

The media's emphasis on the negative aspects of synthetic chemicals may lead one to believe that "natural" products are safer, yet the most potent toxins and carcinogens are of natural origin. There is no requirement, at present, to identify the components of natural products. Since the chemical identities of the ingredients of many natural products are unknown, the safety or risk related to their use is also unknown, and it is practically impossible to prove any association between the use of the product and any adverse health effects which may later appear. Commercial processors of milk, peanut butter, and other foods are required to analyze for aflatoxin and other harmful chemicals, and to keep their presence below specified maximum levels set by the FDA. However, producers of food products classified as "natural," "organic," and "health foods" are not required to analyze for the presence of aflatoxin or other natural contaminants. Inconsistent regulatory policies themselves are an unacceptable risk to our health and environment.

A main classroom building of New York University stands on the site where the Triangle Shirtwaist Company fire occurred in 1911. This fire took the lives of 146 persons, workers who were locked in the cutting and sewing rooms. Locked exit doors, lack of sprinkler systems, and lack of fire-retardant construction materials were common at that time. The workers perished because of the lack of laws to protect their health and safety. Such tragedies now are prevented because of the many laws which have been enacted to protect workers.

In Illinois, also in the early 1900's, the Radium Dial

Company hired young people to paint the faces of clocks with radium to give a glow to the clock faces. Many of these workers engaged in the industry's practice of "tipping" their brushes – a painter licked his brush to form a fine point to paint the fine, decorative numerals, thus ingesting significant amounts of radium. Soon a surprisingly large number of these workers began to die, most from radium-related cancer; however, it was many years before the relationship between the ingestion of radium and the development of cancer was established.

Legislated protection of our health and of our environment is a relatively new benefit for us. Regulations became necessary as a result of the dramatic growth of the chemical industry, catalyzed by the conditions of World War II when most of our natural resources were used to support the war effort. Shortages of food, medicines, and natural fibers, coupled with military needs, prompted research for and development of synthetic rubber tires for vehicles, more durable fabric for uniforms, medicines to combat unfamiliar diseases contracted in foreign lands, and chemical preservatives to permit the transoceanic transport of foods. At the close of World War II, those remarkable technical advances were made available to a public tired of shortages of food, clothing, and nearly all other goods as well. The pharmaceutical, food, consumer product, and chemical industries used the new technologies to improve our nation's standard of living.

Additional technologies brought more new materials and products, a greater proportion of disposable materials, and the coincident generation of new types of residential and industrial wastes. Until recently, we didn't concern ourselves with how these products were made or how the processes of production and disposal impacted our environment or our health. By the time we became aware of the issues of chemical toxicity, hazardous wastes, insufficient testing, and environmental pollution, the giant industries were manufacturing thousands of products and generating tons of waste every day.

An awakened America found the situation to be so complex that they became confused, frustrated, apathetic, and thus eager to accept the "clarifications" presented through the media by special interest groups. We now know there are laws in place to protect our health and our environment, but they are inconsistent in their coverage and enforcement.

We can prevent cancer and other adverse health effects which result from known causes – tobacco use causes 350,000 deaths per year,[3] traffic accidents are the primary cause of death for the 5-35 age group, over-exposure to the sun's radiation results in melanoma and other forms of skin cancer, and over-consumption of saturated fat contributes substantially to heart disease. These are negative factors affecting our well-being that with conscious effort, we can effectively minimize.

Dr. Bruce Ames contends that risks from man-made chemicals are exaggerated and that far greater risks may be generated from natural sources. According to Dr. Ames, the following actions over a lifetime provide far greater risk than drinking four eight-ounce glasses of water per day from the worst-contaminated well (in terms of trichloroethylene concentration) found in California's Silicon Valley: drinking a can of beer a day, swimming for one hour a day in a chlorinated swimming pool, eating a peanut butter sandwich or one raw mushroom per day.[4] Dr. Ames cautions that medical science has much to learn before it can put risk into proper perspective.

Even while we are accumulating knowledge and developing better understandings, we must press for more consistent regulation. Watching our two daughters apply makeup, we feel concern and disappointment with our regulatory system. The cosmetics industry produces more than 25,000 products containing over 8,000 chemical ingredients,[5] but unlike pesticide and pharmaceutical manufacturers, the cosmetic companies are not required to notify the FDA of new products or to tell the Agency of the ingredients included. Even when the FDA receives reports that a cosmetic is unsafe, it can't order the product removed from the market until it is fully reviewed and a court order obtained.

In 1982, the FDA Cosmetics Technology Division had only twenty testers and eight field investigators at their Washington, D.C., headquarters. When they recognize harmful ingredients, they are frequently unable to ban them. For example, many coal-tar dyes are carcinogenic and mutagenic. These dyes are commonly used in hair colorings. Congress specifically exempted coal-tar dyes from the section of the Food, Drug and Cosmetics Act that empowers the FDA to prohibit the use of dangerous "adulterated" substances.

Formaldehyde has been banned from urea-formaldehyde foam insulation because it has been shown to be an animal carcinogen and an irritant. It is still used in eye makeup, nail polish (up to 5 percent) and shampoos. Lead is harmful to our nervous system, but lead acetate is present in hair dyes. The amounts of lead absorbed from the use of these hair dyes are too small to cause measurable lead poisoning in an individual, but by using them, we are putting lead into our sewage systems, knowing that it will eventually get to rivers and even underground aquifers. The small amount of lead discharged from one dying session multiplied by the number of users quickly adds up to hundreds of pounds.

Lanolin, imported from Australia where DDT is still used, is the greasy coating found on sheep's wool which, after refining, is used in cosmetics and ointments. Many chemicals like DDT, PCB's, and dioxin naturally concentrate in fatty matter such as lanolin. A 1988 FDA random analysis of a typical lanolin sold in the United States showed it contained at least fifteen different pesticides.

It is unsatisfactory for any industry to argue that since a mixture of chemicals has a history of safe use, this suggests a lack of toxicity and supports continued use. Benzidine was *safely* used in the dyes and pigments industry until it was found to cause human bladder cancer. Many children were *safely* exposed to the flame retardant TRIS before it was found to be a potent carcinogen and removed from commerce. We should be just as concerned about pesticide residues in cosmetics as we are about pesticide residues in foods.

We must take action as necessary for the protection of our health and for the preservation of our environment and its natural resources. Three of the greatest dangers facing our world community today have a profound impact on the environment. These are: (1) the indiscriminate pollution of our soil, water, and air by major manufacturing and utility industries, (2) continued failure to effectively enforce environmental laws, and (3) our inability to realistically and objectively examine the alternatives to fossil fuels and the burial of wastes. To effect beneficial changes, we must first become knowledgeable about these critical issues affecting us and then actively support politicians, organizations, and programs committed to the protection of us and our environment.

Footnotes

1. P. Preuss, "Summary of *Presentation to the American Association for the Advancement of Science Annual Meeting,*" *Pesticide and Toxic Chemical News*, 24 February 1988, p. 6.
2. United States Department of Agriculture, Washington, DC, 1982.
3. *Smoking and Health: A Report of the Surgeon General*, Department of Health, Education and Welfare Publication No. 79-50066, Washington, DC, 1979.
4. B. N. Ames and others, "Ranking Possible Carcinogenic Hazards," *Science* 236 (1987): 271-279.
5. R. L. Bronaugh and H. I. Maibach, "Cosmetic Safety," *Cosmetic Safety*, J. H. Whittam, Ed. (New York: Marcel Dekker, Inc., 1987), p. 11.

Glossary

acetylsalicylic acid: The active ingredient in aspirin.

acid rain: Raindrops which have the properties of a weak acid.

acute tests: Toxicity tests conducted in animals, in which the animal receives a single dose of the test material, after which the adverse effects are determined for a short time (1-14 days).

aflatoxin B_1: Generally referred to as aflatoxin, this natural carcinogen is the excretory product of the mold, *Aspergillus flavus*.

Alar: A plant growth regulator used by apple growers to slow the ripening of apples on the tree.

amino group: A functional group on a molecule, containing one nitrogen and two hydrogens, NH_2.

aniline: A chemical compound used in the manufacture of dyes and pharmaceuticals.

anticoagulant: A chemical that prevents clotting of the blood.

antifertility activity: Renders an animal infertile.

benzidine: A cancer-causing chemical used in the manufacture of certain dyes.

benzo(a)pyrene: A cancer-causing chemical found in automobile emissions, cigarette smoke, and in effluent from the burning of fossil fuels.

biochemical transformation: The body's ability to convert (transform) one chemical compound into a different compound.

bioconcentration/biomagnification: The accumulation of large amounts of chemicals in body tissues and organs.

biodegradable: A material that is capable of being broken down to smaller pieces by bacteria and natural processes.

botulism: Food-poisoning caused by a chemical toxin (botulin) excreted by the bacterium, *Clostridium botulinum*.

bromadiolone: An effective rodenticide which causes internal bleeding in rodents.

bromdicafum: An effective rodenticide which causes internal bleeding in rodents.

bubonic plague (Black Death): A disease caused by a bacteria, transmitted by rat fleas. In the 1300's, an epidemic in Europe killed one-fourth of the population.

cancer: Uncontrolled malignant tissue growth in animals and humans, which invades surrounding tissue and organs.

carbon dioxide: A colorless, odorless gas formed during fossil-fuel combustion. Chemical formula: CO_2.

carbon monoxide: A highly toxic colorless and odorless gas, which binds to the hemoglobin in red blood cells thus preventing the distribution of oxygen to the vital organs in our body. Chemical formula: CO.

carbon tetrachloride: A volatile liquid which has been used as a cleaning solvent, fumigant, and fire extinguishing fluid. Chemical formula: CCl_4. It is now considered a carcinogen and is infrequently used.

carcinogen: Any chemical that is capable of causing cancer.

cell: The basic unit of the living organism, which can carry out all life functions.

Chernobyl: A city in the Soviet Union in which a nuclear reactor accident occurred.

chlordane: A chlorinated pesticide used to eliminate termites.

chlorination: The addition of chlorine to water as a disinfectant.

chloroform: A volatile liquid, which has been used as an anesthetic. Chemical structure is $CHCl_3$.

chlorofluorocarbons: Chemicals containing chlorine and

fluorine, used in freons as refrigerants, in aerosols, and in fire extinguishers.

cholesterol: A steroid present in all animal tissues and concentrated in the brain, spinal cord, and skin.

chromosomes: Segments of DNA found within the nucleus of cells and containing the genes. Contains the cell management information and hereditary information.

chronic tests: Animal tests in which chemicals are administered daily to animals, for at least one year, to determine the adverse effects of the test chemical.

DDT: Dichlorodiphenyltrichloroethane. An insecticide that bioaccumulates in ecosystems and in animal tissues.

dealkylation: The cleavage of a molecule into two parts by the liver and certain other organs as a part of the detoxification process.

deamination: The removal of an amine group (NH_2) from a molecule as part of the detoxification process.

detoxification: The body process which converts toxic molecules to less toxic molecules and aids in their excretion from the body.

dibromochloropropane: A pesticide now removed from the marketplace which caused sterility in men, and can cause cancer.

DNA: Deoxyribonucleic acid. The molecule that contains the genetic material (the "genes") which is found in the nucleus of all cells.

dose: The amount of a chemical administered to a test animal at one time.

element: The simplest form of a chemical. Carbon, hydrogen, and oxygen are elements.

environment: The external world surrounding a living organism.

enzymes: Proteins found in the body which catalyze chemical reactions in our cells.

estrogens: A group of female hormones, consisting of estradiol, estriol, and estrone.

ethyl alcohol (ethanol): The form of alcohol found in alcoholic beverages.

FIFRA: The Federal Insecticide, Fungicide and Rodenticide Act, which regulates the manufacture, sale, and use of pesticides.

fission: The splitting of an atomic nucleus. This process, which releases large amounts of energy, generates energy in a nuclear reactor.

flame retardants: Chemicals which decrease or eliminate the flammablility of materials on which they are used.

formaldehyde: A strong-smelling gas which is irritating to the lungs and is used to preserve anatomical specimens, in the embalming of bodies, and as a disinfectant.

fossil fuel: Is formed naturally by the decay of living organisms; coal, oil, and natural gas are fossil fuels.

fructose: A sugar found in fruit juices and honey.

fungicide: A material that kills fungi (mildew, mold, etc).

gene: A unit of genetic information located on the chromosomes.

glucose: A common sugar found in fruits, honey, and in blood.

greenhouse effect: The accumulation of pollution gases in the atmosphere traps the sun's energy in the lower atmosphere, and thus increases the temperature of earth's environment.

groundwater: Bodies of water underground into which wells are dug to obtain drinking water.

hazardous wastes: Materials capable of causing harm because of their ignitability, corrosivity, reactivity, or toxicity.

hemoglobin: The red-colored chemical found in red blood cells, which transports oxygen from the lungs to all the body's tissues.

herbicides: Chemicals that kill vegetation (plants).

hormone: A chemical secreted by one organ of the body that has a biological activity in another part of the body.

hydrocarbons: Molecules consisting of only carbon and hydrogen.

hydroxylation: The addition of a hydroxyl group (oxygen-hydrogen) onto a molecule.

indemnification: A guarantee against financial loss.

infertility: Inability to conceive an offspring.

insecticide: A material that kills insects.

landfill: A large open hole in the ground in which waste material is buried.

leachate: The mixture of chemicals and water which forms after water percolates through buried wastes.

Material Safety Data Sheet (MSDS): A document listing the physical properties, toxicity, antidotal treatment for exposure, protective garments to be worn when handling the chemical, and other information about a particular chemical.

maximum tolerated dose (MTD): The highest dose, in an animal toxicity study, which causes significant toxicity but not an excessive rate of mortality.

metabolism: The conversion of chemicals to other chemicals within the body, for cell growth, energy generation, detoxification, excretion, and utilization of nutrients.

methyl alcohol (methanol): Also called wood alcohol. It is used as a solvent and as an antifreeze. It is poisonous and can cause blindness when ingested.

molecule: The smallest unit of a chemical compound.

natural: Formed by nature, not by man.

neurotoxin: A material which causes an adverse effect on the nervous system.

nicotine: A chemical stimulant found in tobacco. It is highly toxic and was used as an insecticide.

nimby: The acronym for "not in my back yard." Activists against the construction of nuclear power plants, waste incinerators, landfills, etc.

nitrosamines: A class of cancer-causing chemicals.

no-observable-effect-level (NOEL): The highest dose in an animal study which causes no observable toxicity.

nuclear energy: Energy formed by splitting atomic nuclei in a nuclear reactor, using radioactive fuel, such as uranium.

Ordram: A herbicide used to kill weeds in rice fields.

organism: An independent unit able to carry on all functions of life, such as a bacteria, animal, or plant.

organophosphate: A class of chemicals containing carbon, hydrogen, oxygen, and phosphorus, some of which are very toxic, while some are effective insecticides.

OSHA: The Occupational Safety and Health Administration. The government agency which enforces safety in the workplace.

ozone: A reactive form of oxygen (O_3), which is irritating to the respiratory system.

pesticide: A material which kills insects, plants, fungi, or rodents. It can be either natural or man-made.

pH: The unit of measurement to describe the acidity or alkalinity of a material.

polycyclic aromatic hydrocarbons: A class of cancer-causing chemicals formed during the burning of fossil fuels.

pyrethroids: A class of insecticides.

pyrogenic: A material that produces heat and can burst into flames.

radon: A harmful, radioactive gas which seeps out of the ground and can cause cancer.

RCRA: The Resource Conservation and Recovery Act. It regulates the transportation, storage, and disposal of hazardous wastes.

registration: The documented permission required by law and issued by the U.S. EPA for the sale, use, and disposal of pesticides. The EPA awards a registration to a pesticide manufacturer after reviewing the chemistry, toxicology, and environmental information supporting the safety of the pesticide, and concluding that the benefits outweigh the risks.

risk: The chance or possibility that an occurrence will happen.

risk assessment: The process by which potential adverse human health effects are characterized and put into perspective.

salicylic acid: A form of the active medication in aspirin.

SAP: Scientific Advisory Panel. It is a panel of distinguished scientists who make recommendations for actions by the EPA based on their independent assessment of health effects data for specific pesticides.

solanine: A naturally occurring toxic chemical found in potatoes.

Superfund: A special fund allocated for the cleanup of buried chemical wastes.

synthetic: Made by man.

teratogen: A material which is able to produce birth defects.

Thalidomide: A teratogen which caused birth defects in humans.

Three Mile Island: The location of a nuclear power plant.

tomatine: A naturally occurring toxin found in tomatoes.

toxicology: The study of the adverse health effects caused by chemicals.

trace elements: Chemical elements which are very infrequently found in the environment.

TRIS: A chemical that was used as a flame retardant. It was shown to cause cancer in animals.

TSCA: Toxic Substances Control Act. It regulates the production, use, and disposal of certain chemicals.

vinyl chloride: A chemical used in the manufacture of plastics (polyvinyl chloride, PVC), which was found to cause cancer in humans.

willow tree: The bark from this tree was the original source of salicylates, the active medication in aspirin.

warfarin: An anticoagulant, commonly used as a rodenticide.

Index